CAMBRIDGE LATIN AMERICAN STUDIES

39

THE STRUGGLE FOR LAND

A POLITICAL ECONOMY OF
THE PIONEER FRONTIER IN BRAZIL
FROM 1930 TO THE PRESENT DAY

For a list of other books in the
Cambridge Latin American Studies series,
please see page 259

THE STRUGGLE FOR LAND

A POLITICAL ECONOMY OF THE PIONEER FRONTIER IN BRAZIL FROM 1930 TO THE PRESENT DAY

JOE FOWERAKER

Lecturer, University of Essex

CAMBRIDGE UNIVERSITY PRESS

Cambridge

London New York New Rochelle

Melbourne Sydney

PUBLISHED BY THE PRESS SYNDICATE OF THE UNIVERSITY OF CAMBRIDGE
The Pitt Building, Trumpington Street, Cambridge, United Kingdom

CAMBRIDGE UNIVERSITY PRESS
The Edinburgh Building, Cambridge CB2 2RU, UK
40 West 20th Street, New York NY 10011–4211, USA
477 Williamstown Road, Port Melbourne, VIC 3207, Australia
Ruiz de Alarcón 13, 28014 Madrid, Spain
Dock House, The Waterfront, Cape Town 8001, South Africa

http://www.cambridge.org

First published 1981
First paperback edition 2002

A catalogue record for this book is available from the British Library

ISBN 0 521 23555 3 hardback
ISBN 0 521 52600 0 paperback

To Lynne

Contents

Maps

Preface

The pioneer frontier in Brazil may appear at first sight to be another of those rather recondite topics, of concern only to academics. A moment's consideration, however, should be enough to recognise its claims to general interest. This frontier pits man against nature, and demands domination of the physical environment, which is the basis of all economic activity. It involves millions of people over several decades in a massive input of human labour and creation of economic wealth. In its advance it also pits man against man and reveals social and political relationships which often remain hidden within the confines of national society. And with its advance into Amazonia it becomes the world's 'last great frontier', and marks the beginning of the end of a chapter in human history.

I first went to the frontier some ten years ago, and was immediately captivated by the atmosphere and the action, and by the courage of the pioneering peasants. I was struck by the vividness of the experience, and the closeness of earth and elements. At a showing of a 'spaghetti' western in Capanema (Paraná) I was hard put to tell actors from audience. But this romantic vision was soon tempered by the everyday violence of the peasants' lives, and their fears and anxieties. This was evidently less a question of the brutal battle with nature than of 'man's inhumanity to man'. And this did not appear as a moral question, or one that could be answered for all men everywhere, but as a question that could only be answered concretely in the context of the frontier. Long after the original commitment to research this question continued present in my mind.

There have been millions of peasants engaged in pushing back the frontier over the years, and it is their lives which compose the drama of the frontier. I have spoken to some of them, and I have seen the conditions in which they live and work. It has been important to me to explain their situation. To this end I have put together the material which forms the empirical base of this book, and constructed the conceptual framework which gives this material meaning. I still do not know whether it is inevitable or not, but the search for meaning often makes analytical arguments abstract, and so seems to remove them from the realm of 'lived experience'.

If this has occurred here, it was no part of my intention. On the contrary, the concepts and categories are intended to cast light on the lives of these people. As you read this book, they are still actively engaged in a struggle for land, which is a struggle for their very survival.

This struggle on the frontier unmasks the true face of Brazilian society (and, by implication, every underdeveloped capitalist society). In the political and economic centres of this society research may reveal what is said; on the frontier it shows what is done. The confrontations and conflicts which characterise the process of frontier expansion create a special perspective on social organisation and political power, which is simply not available at the 'centre'. Once again, a simple presentation of the empirical evidence, while essential, is insufficient to give significance to the insights gained in this study; hence the need for a conceptual construction of the frontier, and its relation to the national society. Finally the book is as much about Brazilian economy and society as it is about the frontier. One cannot be understood without the other.

Within this perspective the argument provides a historical and structural context for understanding the advance of the frontier into the Amazon basin. In contrast to the pioneering movements in Brazil over the past fifty years, this contemporary development has suddenly been seen as 'significant', and has been debated in the mass media of Europe and the United States. There are fears that the 'last great frontier' will devastate the delicate ecological environment of the Amazon, with dire consequences for global equilibrium. In fact, nobody knows at this moment what will happen in the long term. But it is important to understand what is happening now, and why it is happening; and these questions can only be answered by an investigation into the political economy of the pioneer frontier in Brazil in this century.

Although I had no such ambitions when I embarked on this investigation, it seems at least possible that my argument now represents an attempt at a *regional* political economy. This is an animal much cosseted in certain intellectual circles, but rarely sighted; so I cannot really be sure if I have one here or not. But as the idea of a regional political economy grew in my mind, so I worked towards making one. I doubt if it is complete. There were no recent 'models' for what I wanted to do (indeed, 'political economy' as I understood it seemed to have gone out of fashion before the end of the nineteenth century), so I went on learning by doing. The result you will judge for yourselves.

The organisation of the text is very simple, and divides into three parts, which I like to think of, with Aristotle, as the beginning, the middle, and the end. The first part provides an introduction to the pioneer frontier, and the economic background to its expansion and its relation to the national society. In addition the first chapter reviews the main themes of the book, and previews the principal elements of analysis. The second part looks at different aspects of the role of the State in frontier expansion, including the investigation of law, bureaucracy and violence on the frontier. The third part pulls things 'economic' and things 'political' together in an integrated political economy of the frontier, which explains the phenomena of frontier experience at the level of accumulation, and interprets the growth of the national economy and the formation of the Brazilian State through the perspective of the frontier.

Underlying all this is my concern to keep a proper balance between the presentation of the empirical evidence and its exegesis. The idea is to focus attention continually on the object of study, but, at the same time, to develop the way that object is seen and understood. You may be sure that, simple as that idea may seem, it led to anguished hours of debate over the 'order of exposition' of the piece. Evidently this could not simply correspond to the 'order of research': even were such a thing practicably possible, the process of discovery itself can never solve the problems of building a conceptual framework. On the other hand, I did not want to sharpen the distinction between the two to the point of presenting concrete analyses as mere illustrations of a theoretical interpretation. At the end of this debate I reached a compromise: the thematic content and its analytical concepts are briefly stated at the beginning of the book; but the elaboration of the framework of analysis accompanies the presentation of the empirical material, and advances as tensions in the text demand conceptual clarification. In this way the process of 'conceptual construction' only reaches completion in the final chapters, when sufficient empirical material is present to give it consistency.

Much of this material, not to mention insights into the material, was gained through interviews. I have listed the principal informants at the end of the bibliography, but – in contrast to all written sources of information – I have not attempted to tie down their individual contributions in the text. As a good part of this kind of material is simply 'absorbed' I did not find this a feasible task. Nonetheless I owe these people a lot; their contributions 'inform' the argument throughout. More specifically, those parts of the argument which examine the social and political history of particular frontier regions (such as the south of Mato Grosso in Chapter 4, or Paragominas, Pará, in Chapter 6) come close to being vignettes of oral history. But despite the debt, none of them can be held responsible for any of the views expressed in the text.

The field work on which much of this book was based would not have been possible without the financial support of the Department of Education and Science, which provided funds for the pilot study in 1970; the Ford Foundation, which awarded me a Foreign Area Fellowship for 1971–72; and the Nuffield Foundation, which awarded me a Social Science Research Fellowship for 1976–77. I would like to take this opportunity to thank them for their support. I paid a further visit to Brazil in 1973.

During these visits to Brazil I received gracious hospitality and welcome moral support from friends throughout the country. I cannot list everyone who helped me, but I would like to mention Ricardo and Renata Alves Lima, Roberto and Jean Alves Lima, Carlos and Maria José Alves Lima (all of whom treated me as one of the family); Werner Baer, Celso Guimarães, Oswaldo Aranha, Christopher Pearson, Guarací Adeodato, Felipe and Liana Reichstul, Luis Jorge Wernek Vianna, Manoel Berlinck, José de Souza Martins, Octavio Ianni, Eurico and Janete Lima Figueireido, Orde Morton, Machado and Pina, Charles Wood, Marianne Schmink, Roberto Cortés, Jaime Bevilacqua, Amilcar Tupiassu, and, last but not least, Stephen and Gloria Bunker. Moreover, I think I was especially lucky to be affiliated to CEBRAP (Centro Brasileiro de Análise e Planejamento) in 1971–72, and so have the benefit of advice from Chico de Oliveira and Octavio Ianni.

Many of the ideas which finally found their way into the argument were presented and debated at Glasgow (Institute of Latin American Studies, March 1975); Swansea (Society for Latin American Studies conference, April 1975); Oxford (Centre for Latin American Studies, June 1975); London (Institute of Latin American Studies, January 1976); Liverpool (Latin American Centre, March 1976); and Houston (joint Latin American Studies Association and African Studies Association conference, November 1977). I am grateful for these opportunities for criticism and clarification.

I also needed help in putting together the typescript. Professor Bo Särlvik, Chairman of the Department of Government at the University of Essex, answered my calls for help, and Joanne Brunt typed nearly all of the first draft from my dictation at a time when I was *hors de combat*. Professor Terry McCoy, Associate Director of the Centre for Latin American Studies of the University of Florida at Gainesville, was equally supportive, and Kathleen Stipek, Betsy Roberts and especially Lydia Gonzalez cooperated in typing the final version. Paul Stayert drew up the maps. My thanks to them all.

Finally I want to thank all those who have commented on this text, or on my previous attempts to get to grips with the topic, and especially Alan Angell, Malcolm Deas, Peter Flynn, Laurence Whitehead, Ian Roxborough, Ernesto Laclau, Harold Wolpe, Ruben Zamora, Stephen Bunker, Marianne Schmink, and Charles Wood. All of these people helped make this a better book than it might have been, but whatever errors remain are my responsibility. After all this time I could wish that it were a better book than it is, but, for all that, I do not regret it.

Parts of the argument of chapter 9 were previously published in an essay on 'The contemporary peasantry, class and class practice', in Howard Newby (ed), *International Perspectives in Rural Sociology*, John Wiley & Sons, 1978.

Gainesville, 27 December 1979

Glossary of acronyms and abbreviations used in the text

ACAR	Associação de Crédito e Assistência Rural
ACARPA	Associação de Crédito e Assistência Rural do Paraná
ARENA	Aliança Renovadora Nacional
BASA	Banco da Amazônia S.A.
BB	Banco do Brasil
BNDE	Banco Nacional do Desenvolvimento Econômico
BRAVIACO	Companhia Brasileira de Viação e Comércio Ltda.
CANGO	Colônia Agrícola Nacional General Osório
CCEEA	Comitê Coordenador dos Estudos Energéticos da Amazônia
CEFF	Comissão Especial da Faixa de Fronteira
CFP	Comissão do Financiamento de Produção
CIBRAZEM	Companhia Brasileira de Armazenagem
CITLA	Clevelândia Territorial e Agrícola Ltda.
CNA	Confederação Nacional de Agricultura
CODEPAR	Companhia do Desenvolvimento do Paraná
CPRM	Companhia de Pesquisa de Recursos Minerais
CSN	Conselho de Segurança Nacional
CVRD	Companhia Vale do Rio Doce
DGTC	Departamento de Geografia, Terras e Colonização (Paraná)
ELETROBRAS	Centrais Elétricas Brasileiras S.A.
EMBRAER	Empresa Brasileira de Aeronáutica
EMBRATUR	Empresa Brasileira de Turismo
FIDAM	Fundo para Investimentos Privados no Desenvolvimento da Amazônia
FINAM	Fundo de Investimentos da Amazônia
FINOR	Fundo do Investimentos do Nordeste
FISET	Fundo de Investimentos Setoriais
FPCI	Fundação Paranaense de Colonização e Imigração
FUNRURAL	Fundo de Assistência ao Trabalhador Rural

GETSOP	Grupo Executivo das Terras do Sud-oeste do Paraná
IBDF	Instituto Brasileiro do Desenvolvimento Florestal
IBGE	Instituto Brasileiro de Geografia e Estatística
IBRA	Instituto Brasileiro de Reforma Agrária
ICM	Impôsto de Circulação de Mercadorias
IGRA	Instituto Gaúcho de Reforma Agrária
INCRA	Instituto Nacional de Colonização e Reforma Agrária
INDA	Instituto Nacional do Desenvolvimento Agrário
INIC	Instituto Nacional de Imigração e Colonização
INP	Instituto Nacional do Pinho
IPEA	Instituto de Pesquisa em Economia Aplicada
ITERPA	Instituto de Terras do Pará
ITR	Impôsto Territorial Rural
MARIPA	Industrial Madeireira e Colonizadora Rio Paraná S.A.
MDB	Movimento Democratico Brasileiro
MINTER	Ministerio do Interior
PDA	Plano do Desenvolvimento da Amazônia
PETROBRAS	Petróleo Brasileiro S.A.
PIC	Projeto Integrado de Colonização
PIN	Plano de Integração Nacional
PND	Plano Nacional do Desenvolvimento
PRODOESTE	Programa do Desenvolvimento do Centro-Oeste
PRORURAL	Programa de Assistência ao Trabalhador Rural
PROTERRA	Programa de Redistribuição de Terras e de Estímulo à Agroindústria do Norte e Nordeste
PSD	Partido Social Democrática
PTB	Partido Trabalhista Brasileiro
RADAM	Projeto Radar da Amazônia
RTZ	Rio Tinto Zinc
SAGRI	Secretaría da Agricultura (Pará)
SEIPU	Superintendência das Emprêsas Incorporadas ao Patrimônio da União
SPVEA	Superintendência do Plano de Valorização da Amazônia
STF	Supremo Tribunal Federal
SUDAM	Superintendência do Desenvolvimento da Amazônia

SUDECO	Superintendência do Desenvolvimento do Centro-Oeste
SUDENE	Superintendência do Desenvolvimento do Nordeste
SUDEPE	Superintendência do Desenvolvimento da Pesca
SUFRAMA	Superintendência da Zona Franca de Manaus
SUPRA	Superintendência da Reforma Agrária
UDN	União Democrática Nacional
USAID	United States Agency for International Development

Glossary of words and phrases in Portuguese used in the text

alqueire a measure of land, varying in size according to geographical zone; the *alqueire* referred to here is the *alqueire paulista*, equivalent to 2.42 hectares (while one hectare = 10,000 m^2 = 2.471 acres).

apossamento occupation and appropriation of land by individual initiative; the act or process of establishing *posse*

autarquia semi-autonomous Federal State department or 'bureau'; these are separate from but subordinate to the diverse Ministries and have independent resources and decision-making capacity.

aviador intermediary (warehouse or individual) which supplies *seringalista* with merchandise for his operations and receives on consignment the rubber collected from the area under his control.

aviamento the system of rubber collection and export which links commercial and industrial capital to the rubber collectors by a series of debt relations through the *aviador* and *seringalista*; the system is characterised by its highly exploitative nature and the direct coercion applied to the collectors.

bandeirante member of armed band of early explorers in Brazil.

barracão the company store, or the store from which the *seringalista* supplies his rubber collectors.

bóia-fría lit. cold grub; rural labourer who lives in a small town or on the outskirts of a larger one, and who is contracted in gangs on a daily or piece-work basis for work in the fields.

borracha rubber

caboclo a Brazilian 'backwoodsman' (originally a white and Indian half-breed) who practises a very rudimentary agriculture.

cangaçeiro — bandit; outlaw. Usually refers to large bands operating in the North-East of Brazil at the end of the last century and beginning of this, and enjoying relative immunity from the law.

cassação — deprivation of political rights by the State (used extensively after 1964 as a measure of political repression).

caúcho — a gum-tree from which the wild rubber known commercially as 'caucho' is obtained.

centro — small settlement which lies (paradoxically) far into the forest, at the furthest extension of the frontier, or where rubber-collectors construct their camp.

colonização Coca-Cola — a 'fake' colonisation, where the plans on paper are not carried out on the ground; the overtones are obvious.

colonização mansa — an ordered or 'domestic' colonisation, which is relatively unmarked by legal disputes and social conflict.

comodato — the right to plant subsistence or cash crops for a limited period in return for the clearing of land (compare *troca pela forma*)

conta corrente — 'current account'; used to describe the indebtedness which binds the peasant to the middleman.

coronel — backlands boss; landowner of the interior who holds political and sometimes military power in his locality or region

coronelismo — system of political power, and control of peasant populations, based on the *coronel*.

cuiabano — native or inhabitant of Cuiabá (Mato Grosso).

cultura efectiva e morada habitual — stock legal terminology for living on the land and farming it.

derruba — felling of trees (to clear the land).

desbravador de mato — trail-blazer; pioneer who arrives first to the frontier.

direito de posse — the legal claim to public land achieved through its occupation and appropriation.

documentação fria — forged documents or forged title to land.

drogas do sertão — medicinal herbs and fruits found wild in the interior.

empreiteiro — a contractor of labour; usually one who hires gangs for work on the large estates of the interior, especially in Amazônia.

engenho	sugar mill and plantation complex
faixa de fronteira	frontier strip, usually defined as 66 kilometers broad, bordering the continental political boundary.
gato	lit. cat; colloquial for *empreiteiro.*
grilagem	land grabbing, especially by use of invalid documents or forgery of titles.
grileiro	one who practises *grilagem.*
indústria de posse	systematic trade in *direitos de posse* which profits from legal disputes over ownership of land in frontier regions.
(por) interêsse social	in the public interest; has come to mean 'where there is social unrest or revolt' in cases of land expropriation.
interventor	the 'temporary' governors of local states appointed by Vargas following the Revolution of 1930 (the practice continued until the end of the Estado Novo).
intrusagem	illegal occupation of property in land.
intruso	trespasser; one who occupies the land without legal title, or any other right.
invasor	trespasser; usually used in a stronger sense than that of *intruso*, as one who deliberately flouts the legal title to the land he takes over.
invernista	one who provides pasture for fattening cattle (originally this would be during the 'winter').
jagunço	hired strong-arm man or bandit; thug.
jôgos de advogados	colloquial expression for the complex legal manoeuvres which accompany litigation over the right to land.
lavrador de asfalto	lit. asphalt farmer; used ironically to denote a city-bound title-holder or speculator.
macho	manly, strong, 'tough'.
mato	woods, forest, jungle; brush, undergrowth.
mata-páu	lit. strangler-tree; colloquial for gun-men or killers in pay or service of the police.
município	municipality (division of local government in Brazil corresponding roughly to a county), which is divided in turn into *distritos* (districts), *vilas* (towns) and *linhas* (rows of dwellings comprising small farming settlements).
na empreitada	work done by contract in gangs (with overtones of hard, unremitting labour).

na fôlha 'on the sheet' or 'on the register'; refers to fixing of the price for the sale of goods by the peasant to the middleman before the harvest is in (the peasant's indebtedness prevents him from getting the full market price for his produce).

paranaense of the state of Paraná; native of this state.

parcería share-cropping.

peões peons, peasants on landed estate or enterprise who are controlled, variously, by coercion, debt relations, personal obligations.

posse occupation of land which implies a claim on that land.

povoado village, or settlement; usually behind the furthest extension of the frontier, but where commercial relations are established; or where the rubber boss manages his operations (also called *beira* or *margem*).

procuração falsa bogus proxy; refers to the titling of land to fictitious persons who then 'give' power of attorney to economic interest groups or land speculators.

registro land register of a judicial district.

safrista one who fattens pigs on the maize at the time of the *safra* (harvest).

seringalista rubber boss; controller of area of forest where rubber trees are found, who oversees labour of *seringueiros*.

seringueiro rubber-gatherer; rubber-tapper.

sesmaría land grant in colonial Brazil.

sindicato labour union.

sistema de bugre rudimentary farming techniques employed by the Indian or half-caste populations of the interior (compare *caboclo*).

tenentista movement a political movement for the reform of the Brazilian political system, originating among the junior officers of the army (*tenentes*) in the 1920s.

terras devolutas unclaimed public lands.

troca pela forma contract between landholder and labourer or peasant, whereby the land is cleared in return for the right to plant cash or subsistence crops for a limited period, usually one to three years (compare *comodato*).

volante casual rural labourer who, once brought to the interior on a 'contract', must now seek work of any kind to survive (used especially of North-easterners in the Amazon region. Compare *bóia-fría* and *empreiteiro*).

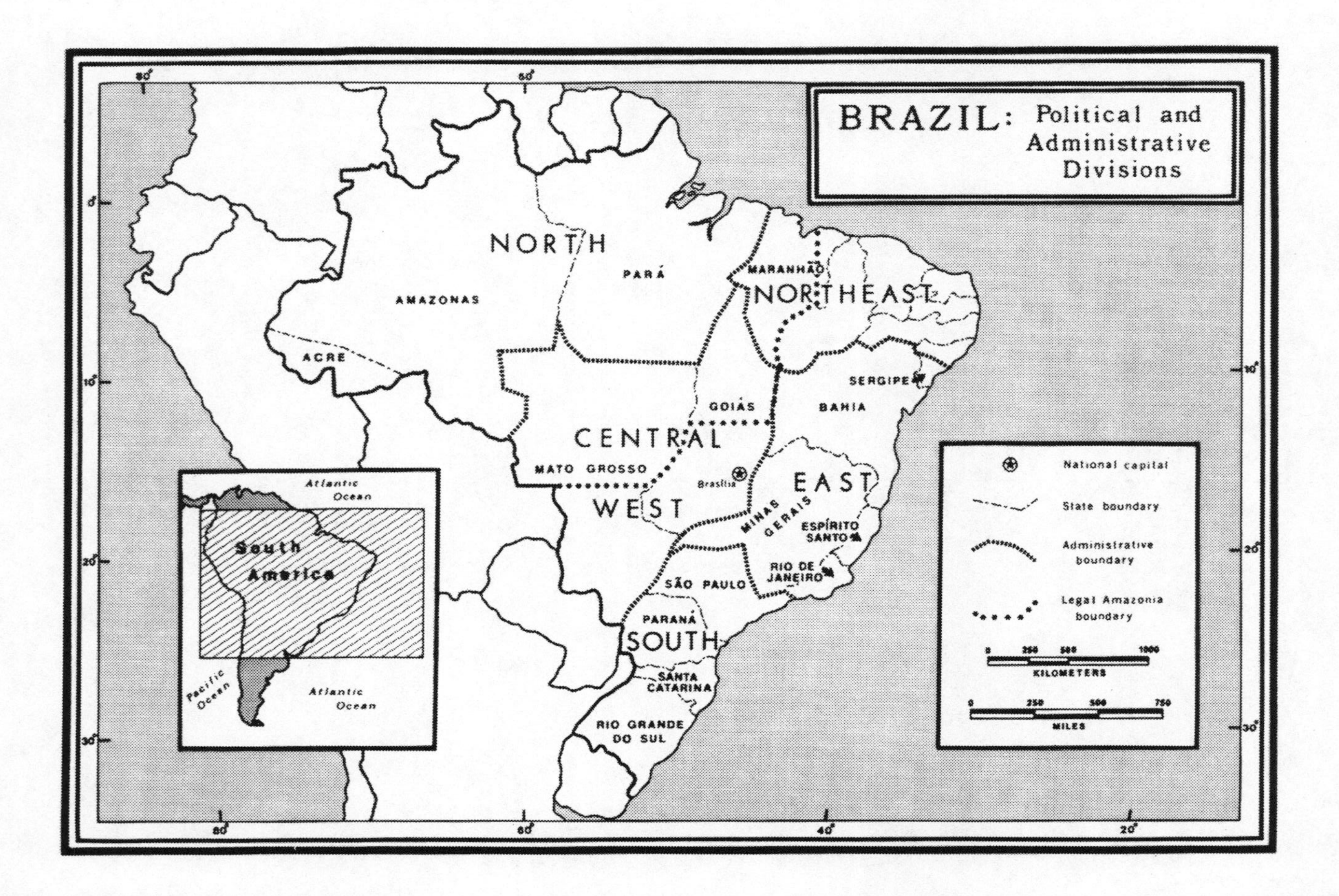
BRAZIL: Political and Administrative Divisions
NORTH
AMAZONAS
PARÁ
ACRE
MARANHÃO
NORTHEAST
SERGIPE
BAHIA
GOIÁS
CENTRAL WEST
MATO GROSSO
Brasília
EAST
MINAS GERAIS
ESPÍRITO SANTO
RIO DE JANEIRO
SÃO PAULO
PARANÁ
SOUTH
SANTA CATARINA
RIO GRANDE DO SUL
National capital
State boundary
Administrative boundary
Legal Amazonia boundary
KILOMETERS
MILES
South America
Atlantic Ocean
Pacific Ocean
Atlantic Ocean

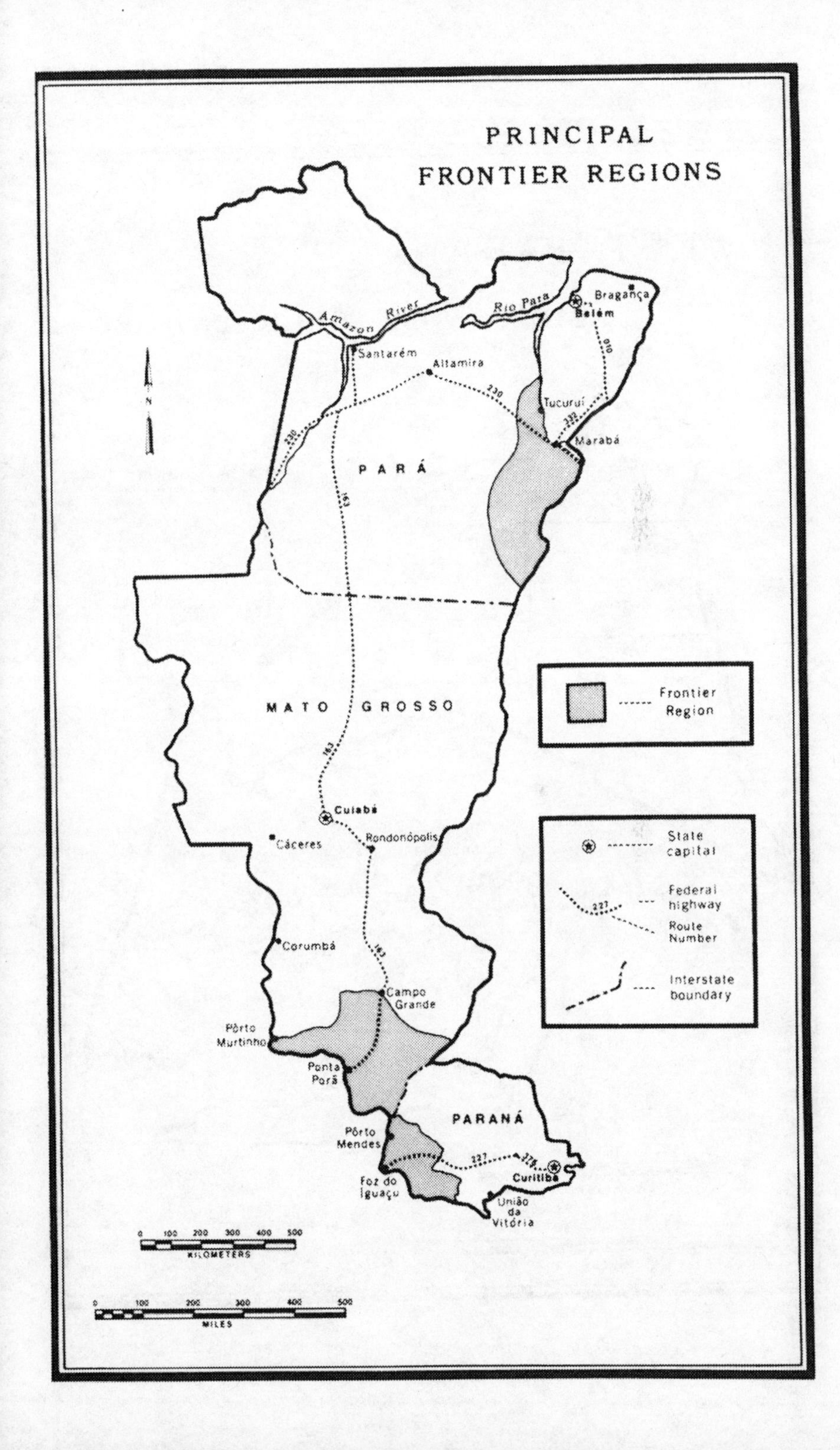

PRINCIPAL
FRONTIER REGIONS
Amazon River
Rio Pará
Bragança
Belém
010
Santarém
Altamira
230
Tucuruí
322
Marabá
PARÁ
163
MATO GROSSO
Frontier Region
Cuiabá
Cáceres
Rondonópolis
Corumbá
State capital
Federal highway
227
Route Number
Interstate boundary
Campo Grande
Pôrto Murtinho
Ponta Porã
PARANÁ
Pôrto Mendes
Foz do Iguaçu
277
Curitiba
União da Vitória
0 100 200 300 400 500
KILOMETERS
MILES

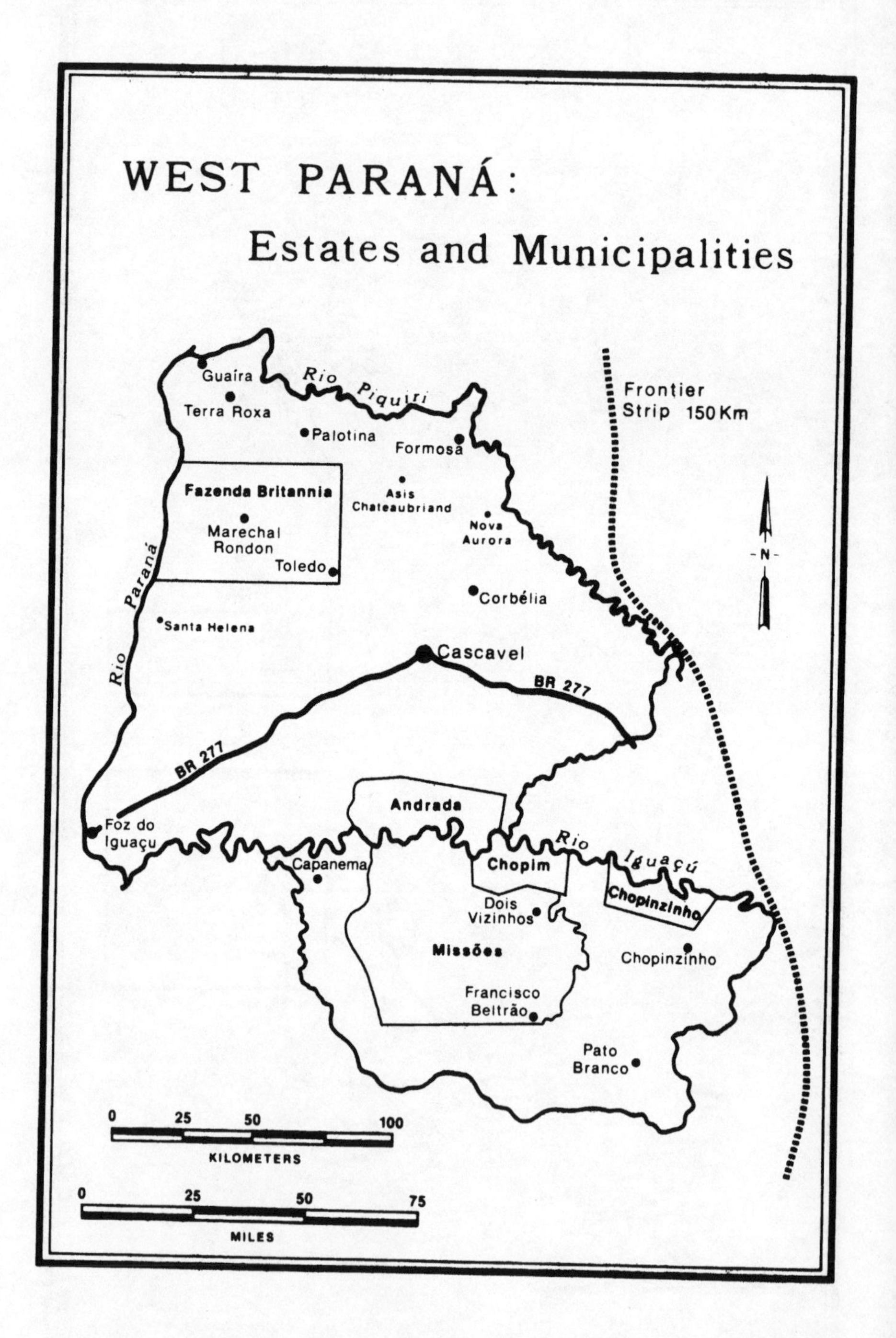

WEST PARANÁ:
Estates and Municipalities
Guaíra
Terra Roxa
Rio Piquiri
Palotina
Formosa
Frontier Strip 150Km
Fazenda Britannia
Asis Chateaubriand
Marechal Rondon
Nova Aurora
Toledo
Rio Paraná
Corbélia
Santa Helena
Cascavel
BR 277
BR 277
N
Andrada
Foz do Iguaçu
Rio Iguaçú
Capanema
Chopim
Chopinzinho
Dois Vizinhos
Missões
Chopinzinho
Francisco Beltrão
Pato Branco
0 25 50 100
KILOMETERS
0 25 50 75
MILES

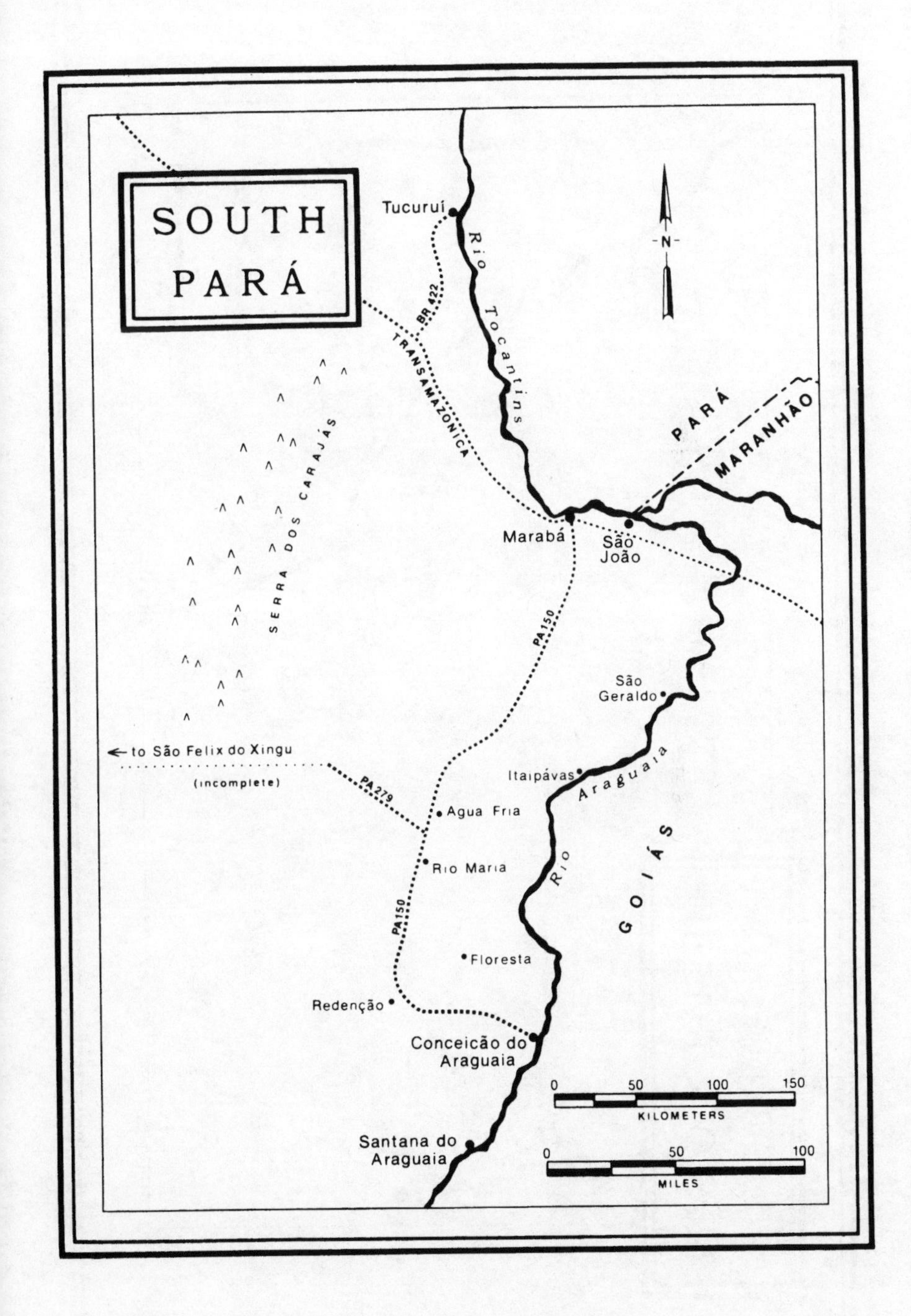
SOUTH
PARÁ
Tucuruí
Rio Tocantins
BR 422
TRANSAMAZONICA
SERRA DOS CARAJÁS
PARÁ
MARANHÃO
Marabá
São João
PA150
São Geraldo
← to São Felix do Xingu
(incomplete)
PA 279
Itaipávas
Rio Araguaia
Agua Fria
Rio Maria
GOIÁS
Floresta
Redenção
Conceicão do Araguaia
Santana do Araguaia
0 50 100 150
KILOMETERS
0 50 100
MILES
N

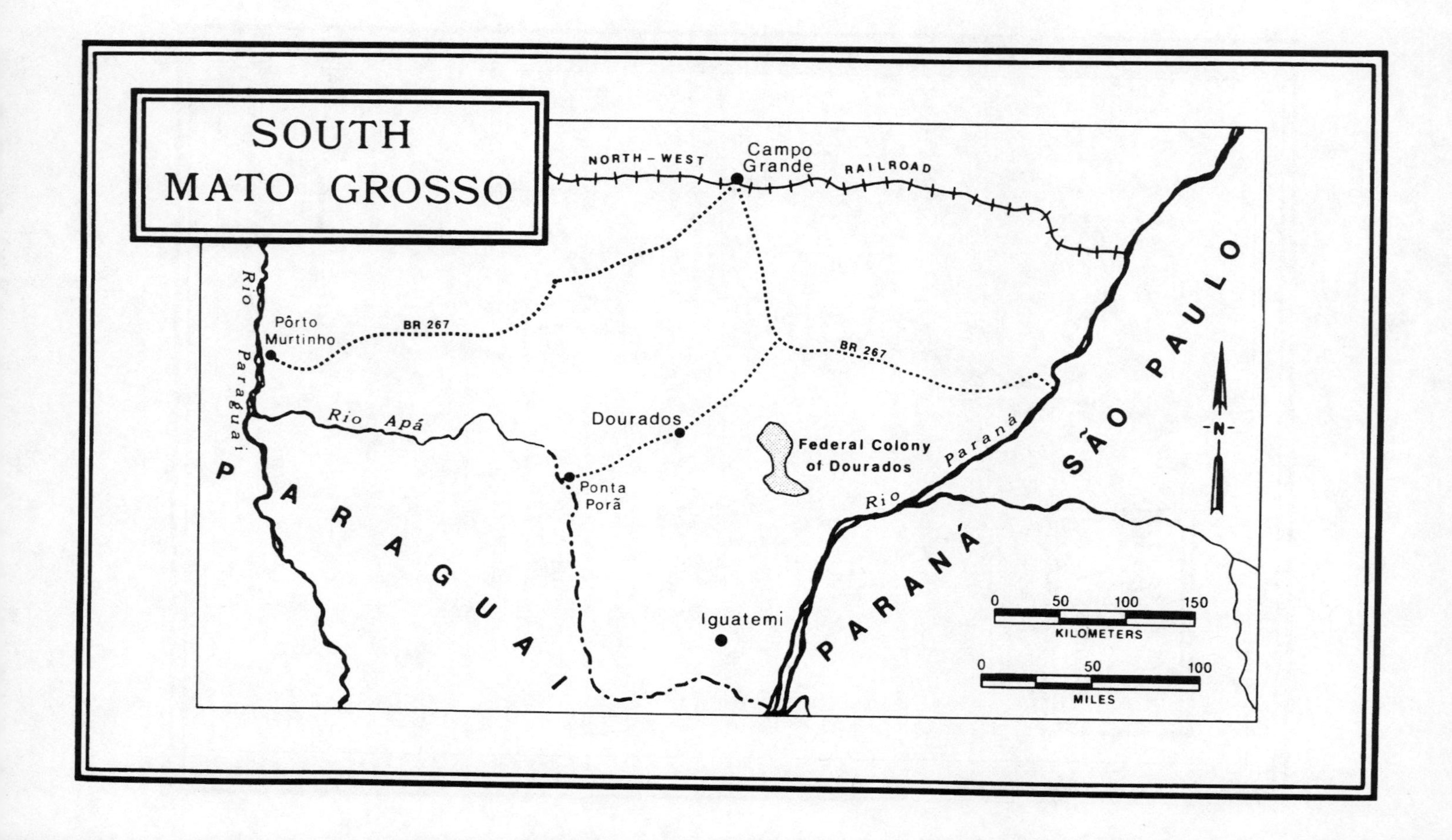
SOUTH
MATO GROSSO
NORTH - WEST
RAILROAD
Campo
Grande
BR 267
BR 267
Pôrto
Murtinho
Rio
Paraguai
Rio Apá
Dourados
Ponta
Porã
Federal Colony
of Dourados
Rio Paraná
Iguatemi
PARAGUAI
PARANÁ
SÃO PAULO
N
0
50
100
150
KILOMETERS
0
50
100
MILES

PART I

The pioneer frontier

I

The pioneer frontier: political violence and the peasantry

The pioneer frontier: specificity and generality

The aim of this book is to reach an understanding of the pioneer frontier in Brazil. The object of study is conceived as the particular process of frontier expansion occurring in the country over the last half century. This concept of frontier in no way corresponds to the so-called cyclical character of economic growth and occupation of land in Brazil. There is, therefore, no intention here of following precedent (Normano 1935; Castro 1969) and presenting the entire economic history of the country in terms of its 'frontier' experiences. These growth cycles have been observed to follow the economic booms in different products for export to the world market – such as sugar, gold, coffee and rubber – and have depended on new demands arising within that expanding market over the centuries (Prado 1962a; Furtado 1963). The pioneer frontier, on the contrary, has expanded in response to the demands of the national market and in function of economic accumulation within the national economy since 1930.

It is to be expected that as the concept of frontier gains currency it will lose content. There is already an account which assimilates most of Latin American history to the idea of 'frontier' (Hennessey 1978). So it must be clear at the outset that the pioneer frontier is a process of occupation of new lands which is historically specific. The period of the process corresponds to the period of Brazil's most rapid rates of industrialisation and urbanisation, and begins at the moment when the Brazilian economy, for the first time in its history, experiences a large labour surplus (arguments which are taken up in Chapter 3). Just as the national economy grows 'in depth' in the industrial and financial centres so it grows 'in breadth' through the extension of the pioneer frontier. The frontier expresses not any and all economic activities directed to the world market, but the particular activity which integrates unexplored regions into the national economy. The process is propelled by the forces and contradictions of this economy.

Where the history of this economy is viewed 'cyclically' the cycles represent the rise and fall of economic activity in general in one or other region of the country at different times. The process of the pioneer frontier is also viewed cyclically but with the crucial difference that here the cycle is one of accumulation and appropriation of a surplus which can occur simultaneously on diverse frontiers throughout the country. In the review of the historical 'cycles' it seems not to matter whether the export boom was based on slave, servile or 'free' labour; in other words the analysis is not focussed on the mode of production, but advanced at the level of the world market. In the case of the pioneer frontier cycle however, it is important to the analysis that the national economy is clearly capitalist. While the mechanisms of accumulation on the frontier may not themselves be capitalist, the surplus is expropriated not only by speculatory and commercial but also by industrial capital. Within this perspective the frontier cycle is primarily determined by the capitalist social relations which dominate the social formation, and is achieved through a wide range of political, legal and ideological interventions by a particular form of the capitalist State.

These assertions are intended to bring the principal premises of the argument into view. They do not deny the historical diversity of the pioneer frontier experiences in Brazil, but nevertheless maintain that the different frontiers with their particular features are all part of the one secular process of the occupation of the land in modern Brazil, and, as such, are all equally the historical result of a similar set of determinations. Until now, the diversity of the experiences appears to have discouraged attempts at a general approach to the question of the frontier. The classic accounts by Monbeig (1952) and Roche (1959 and 1968) eschewed the problem of generality, and were confined to a careful analysis of particular experiences; contemporary accounts on the other hand by Velho (1972) and Martins (1975) attempt to develop typologies of the frontier, so dividing it into distinct phenomena (the 'expanding frontier'; the 'demographic frontier') which seem to deny the one process. The premises of the present approach to the problem, however, not only insist on what it is that makes these frontiers distinct but also what it is they have in common; the aim is to demonstrate both the specificity and generality of the pioneer frontier.

These opening remarks take on relevance in the face of the empirical reality. While most frontiers move onto virgin lands at the limits of penetration into the interior, others may 're-discover' regions which had previously known occupation during an economic boom (like the Baixo Rio Doce or Espirito Santo), or enter areas which had previously been by-passed for lack of economic attraction at the time. The virgin lands they occupy may be covered with tropical jungle (Pará), pine forest (Paraná), savanna and scrub (Goiás and Mato Grosso), or natural pastures. The soil on the frontier may be sandy and acid, or rich and fertile. The dominant economic activity will vary from casual to organised extractive activity, from small-scale farming to large-scale cattle raising; and the level of technology employed will range from the rudimentary combination of land and labour in 'slash and burn' agriculture to highly capitalised agro-industrial enterprises. Finally, the frontier may be almost entirely isolated or may be integrated by asphalt roads and a developed marketing network into regional and national economies.

In addition to differences of this order the final pattern of settlement on the frontier may vary radically. In some cases the frontier will continue to absorb a large flux of migrants over a long period of time, and the land will be occupied by small farmers engaged in regular agricultural production. Such settlement has occurred in the north-east of Rio Grande do Sul, the west of Santa Catarina, the west and north of Paraná, the south of Mato Grosso, areas of the centre-west of São Paulo, the south of Goiás, the valley of the river Doce in Minas Gerais, a large part of Espirito Santo, the west of Maranhâo, and today in areas of Pará and Rondônia. Elsewhere the migrants to the frontier have been lucky to farm the land for two or three years – if at all. Increasingly in Brazil, and especially in Amazônia, land on the frontier is taken over by large holdings and large enterprise, dedicated more often than not to cattle-raising. In these cases the cattle grow fat on fine pastures, while the people go hungry.

Despite this diversity of experience, it is possible to generalise about the pioneer frontier by means of a political economy which locates it in the context of the national economy and society, and explains the expansion at the level of accumulation. For instance, it may be relatively and increasingly rare for pioneers on the frontier to stay on the land and farm it, but nevertheless it is they who repeatedly take on the task of clearing the land and by their labour create value; it may have been yet rarer in the past for those who did stay on the land to become healthy homesteaders like those of North America, but nevertheless they produced values for the national market. The problem in approaching the difference between 'farming' and 'cattle' frontiers, on the one hand, or the regime of minifundio, on the other, is to discover the production and market relations which achieve the appropriation of this value and so establish their place in the cycle of accumulation on the frontier. Some of the apparent differences in these diverse experiences can be assimilated by referring them to the different production and market relations existing at different 'stages' of the accumulation cycle (which is the argument of Chapter 2). Some of the major similarities, on the other hand, can only be understood once it is discovered that much of the accumulation takes place outside and beyond production and market relations, by means of a form of primitive accumulation (which is the theoretical position adopted in Chapter 8).

Thus the argument of the book makes bold to talk of the pioneer frontier in general, but – it will be seen – this same argument refers repeatedly to three particular frontiers. I can best explain this by speaking personally for a paragraph. The argument is made general because I wished to construct a political economy of the frontier; but the references are particular because there exist very few regional studies which are useful to this effort, and I have drawn heavily on my own field work in the west of Paraná, south of Mato Grosso and south of Pará. The case material from these regions forms a large part of the empirical base of the study, and I submit that without this field work the study would have suffered from a lack of documentation. But while it may be clear why I present case material of this kind at all, it is equally important to explain why I went to *these* frontiers – among the many possible options for research. It is not only necessary but will be useful to defend my choice of case material, as in doing so I can conveniently preview the principal organisational themes of the book.

Paraná, Mato Grosso and Pará: the question of periodisation

It will do no harm to begin with the obvious. The very size and regional diversity of Brazil has always made generalisations about the country precarious. As the investigations of the frontiers in question were designed to illumine a political economy of the frontier at the level of national society, a minimum condition of the choice dictated that these frontiers lie in different regions of the country. The west of Paraná and the south of Mato Grosso lie in the south and centre-west, respectively, while the south of Pará is situated in the north (below the mouth of the Amazon). This same geographical spread further implies that the ecological environment (soils, climate, vegetation) of the frontier varies widely from one case to another. At the same time these frontiers demonstrate different patterns of occupation, deriving from the different historical conditions of control and appropriation of the land; from the different impact of extractive industries; from the intensity and timing of migration onto the frontier. These patterns of occupation are examined in detail in the following chapter, but, by way of illustration, the west of Paraná sees a heavy influx of smallholders onto lands covered by virgin pine forest; the south of Mato Grosso sees a similar experience but within a region of traditionally large landholdings owned by big companies and local political bosses; and in Pará the traditional leasing of land for extractive industry is submerged by the competition between big capital and small peasants for this land, in the wake of the Federal State road-building programme.

While all these varying conditions were pertinent, the central consideration governing the choice of case material was that frontier expansion into these three regions occurred at different *periods* (within the overall period of the expansion of the pioneer frontier, as suggested above). So, very approximately, the frontier in the west of Paraná expanded most rapidly between 1945 and 1970, that of the south of Mato Grosso between 1955 and 1975, and that of the south of Pará from 1965 to the present. Moreover, this 'lag' between Paraná and Pará is not merely coincidental insofar as it is the 'closing' of the frontiers in the south which contributes to propel their expansion into the Amazon region of the north (Katzman 1977a). The significance of the difference in period is simple but far reaching. It allows the investigation of the *changes* in the process of the pioneer frontier which reflect more or less directly the changes occurring at the level of the national political economy.

In effect, a large part of the historiography insists, in the first place, on the changes made by the 'Revolution' of 1964 in the national political and administrative structures, which are demonstrated most strikingly by the greatly increased power and penetration of the Federal State apparatuses, and by the consequent decline in the political power and autonomy of the local state administrations (such as those of Paraná, Mato Grosso and Pará). Moreover, there are those who see these changes at the political level as necessary and logical adjustments to the new social realities created by the changing structure of the Brazilian economy in the modern period: the increasing preponderance of foreign monopoly capital in the manufacturing sector and the rapid, if highly selective, capitalisation of the countryside. Evidently, things 'economic' and things 'political' cannot be divorced in this context, and the greater number, range and autonomy of the Federal regional and sectoral agencies represents both a response to the greater complexities and contradictions of the economy and an extension of State participation in the process of accumulation (growth of State manufacturing and mining sectors). But the main point emerges clearly: the period of the pioneer frontier is not a homogeneous period at the level of the national political economy, and this is necessarily reflected in the process of frontier expansion.

There is no doubt that changes at the level of national economy and national polity bring changes to the pioneer frontier. This is seen especially in the impact of the political on the economic. For example, the investigation of frontiers in different periods allows a demonstration of the greater relative autonomy of the local state administrations in the period before 1964 (in Mato Grosso and Paraná) and the progressive, but certainly not total, loss of such autonomy after 1964 (in the cases of Pará and Paraná). These changes at the political level impinge directly upon the process of accumulation on the frontier by altering in some degree the relative participation of local and national dominant classes in the appropriation of the surplus and by concentrating bureaucratic intervention in this process at the Federal level. None of this is yet meant to raise the question of the 'final' determination of these changes; the relation between local states and the Federal State is important partly because it is itself visible, and partly because it provides a clear view of how accumulation on the frontier is achieved. For these reasons it emerges as a major theme of the book, and the difference in frontier periods captures changes in this relation, and hence gains insights into the changing structure of State in Brazil.

If this was all there was to be said on the question of period, then it would have been analytically advantageous to present the case material chronologically, so that all was known about Paraná, for example, before broaching the case of Pará. Such an approach would certainly have been more considerate of the reader. As it is, the exposition of the material switches back and forth between the different cases in a way which must occasionally be very demanding of the reader. In other words, the argument rejects the difference in period of these experiences as the principal element of their comparison and compares the different frontiers directly despite the difference in period. This strategy is preferred in the firm conviction that whatever the changes symbolised by the Revolution of 1964, far more significant in the political economy of both nation and frontier over the period are their *continuities*. The choice of frontiers at different periods allows an appraisal of the changes, but their direct comparison is designed to demonstrate and emphasise the continuities.

Within this perspective the changes at the economic level in the degree of concentration of capital in the economy at large, and in the degree and rate of capitalisation of production in the countryside, are precisely changes in degree which do not transform the dominant social relations of production in the economy, nor the primary economic determination of frontier expansion as a cycle of accumulation. It is true that monopoly capital domination of the economy may actually accelerate this cycle of accumulation, and that, increasingly, capitalist social relations may emerge towards the end of each cycle, but, equally, the form of primitive accumulation peculiar to the pioneer frontier continues, and the role of the frontier in reproducing the conditions of accumulation in the Brazilian countryside remains essentially the same.

There are at least two ways in which this position might be modified. In the first place, as indicated above, the relative participation of different fractions of capital (local, national, monopoly) in the appropriation of surplus from the frontier may change over time, and consequently so may the application of the surplus. However, there is evidence to suggest that changes of this sort, if they reflect changes in the national political economy, are also directly related to the different 'stages' of the accumulation cycle on the frontier (as is argued in Chapters 2, 6 and 8). In the second place, at a yet higher level of abstraction, the frontier is seen as reproducing the conditions of accumulation in the countryside, which, traditionally, has meant the reproduction of a highly concentrated pattern of land-ownership, and the extension, partly by means of this monopoly in land, of a sub-capitalist economic environment. In this conceptual paradigm the 'national society' is seen as a social formation characterised by the *articulation* of different modes of production, where the capitalist mode is dominant. Let it be said immediately that such concepts demand systematic historical specification before they can become useful tools for analysis – and this is part of the burden of the book. The point to note here is that the role of the frontier in reproducing the articulation remains the same, but the form of the articulation may change with the (selective) emergence of capitalist social relations in the countryside.

Within this same perspective the changes in the lines of command at the institutional level and the degree and depth of the political penetration of the frontier are again, precisely, changes of degree which do not alter, in any essential way, the kinds of political and ideological intervention occurring on the frontier. The institutional initiatives may change, as a result, for instance, of the changing bureaucratic balance between local state and Federal State; but, in so far as political and ideological intervention on the frontier contributes to achieve an appropriation of surplus and to complete the cycle of accumulation, then they constitute *forms of mediation* which remain the same over time. Mediation is defined here, in its broadest sense, as the process of institutionalisation of class struggle, which is always a struggle for social surplus, and, on the frontier, is also a struggle for land.

The dimensions of this struggle are discussed below. The characteristic mediations of the struggle on the frontier are law, bureaucracy and violence. In this connection, the strategy in the presentation of case material is two-fold: certain material appears recurrently in the argument, but in so far as it demonstrates the operation of different characteristic forms of mediation its relevance alters as the argument advances; at the same time the comparison of different frontiers demonstrates the continuities of these characteristic forms over time. The broad contention here is that the cycle of accumulation of the frontier cannot be understood without a clear conception of the question of mediation; for this reason analysis of the place of law, bureaucracy and violence in the struggle on the frontier occupies much of the central part of the book.

Overall, and taking things 'economic' and things 'political' together, the choice and order of presentation of the empirical data on which the argument is based should now be comprehensible. The greatest weight is given to the case of Paraná, because it so clearly straddles the 'divide' of the Revolution of 1964, and so provides material for demonstrating both changes and continuities. The case of Pará is the most 'modern', and is designed to provide insight into what is happening contemporarily in Amazônia. Mato Grosso receives least attention, but should not for that reason be ignored: a proper reading of this case will correct possible misinterpretations of the presence of large capital and massive Federal intervention in Amazônia as being 'new' phenomena in the political economy of the pioneer frontier. In Mato Grosso the Laranjeiras company monopolised huge tracts of land and kept the migrants off it; the Federal State intervened both economically with a large infrastructure project (the Northwest Railroad), and politically with the large-scale appropriation of private and local state land (the Federal Territory of Ponta Porã) and all this occurred decades before the 'Revolution' of 1964.

The presentation of a political economy

However far the 'economic' and the 'political' are separated for the purposes of analysis and presentation, they must be understood as constituting one, indivisible social process. At the level of general theory Coletti (1972) has argued exhaustively that the different 'instances' of society – economic, political and ideological – are but heuristic devices, and that *social* relations of production must be understood as simultaneously ideological and political. In the case of the pioneer frontier it was stated that its expansion moves through a cycle of accumulation which is determined economically but which is achieved through different forms of political and ideological mediation. Only after locating law, bureaucracy and violence within this cycle of accumulation, that is at the economic level, can the frontier *process* be understood. Moreover, this is specifically necessary in the case of the frontier where one major form of accumulation – primitive accumulation – takes place largely outside relations of production as such, and through the legal and political intervention of the State.

These remarks go some way to explain the logic of the 'order of exposition' of the analysis – which does not follow the logic of the 'orders of determination'. In other words, analysis at the level of economic accumulation (the 'primary' determination) does not always precede analysis at the level of legal and political mediation (the 'secondary' determination) but often follows it. This is partly because, as suggested above, things are not so simple as these 'orders of determination' may imply. But it anyway makes sense to discuss the political mediations first if they are necessary elements in the investigation of the reproduction of social relations of appropriation. The former are, after all, observable, and susceptible to analysis at the political level, using 'middle order' concepts; while analysis of the latter may require a complete conceptual framework capable of integrating the 'political' and the 'economic' (and this analysis is only reached therefore in the closing chapters of the book).

The order of exposition suggested here, and followed in the book, has further advantages in the construction of a political economy. On the one hand, this approach will reveal the struggle of social classes and social forces on the ground (and their specific interests in and response to the range of political and legal interventions) before posing the question of accumulation, and in this way avoid a mechanistic or 'economicistic' determinism. On the other hand, it will explore the internal structure and contradictions of the State (and their relation to particular economic interests and fractions of capital) before this is inserted as the political instance of a determinate social formation, so avoiding the temptation to fetichise the State as being monolithic, or as having some 'reason of State' which is 'above' and 'beyond' the political process of class struggle.

In this connection it should be noted in passing that for the greater part of the book the traditional usage of Brazilian social scientists and others is adopted in distinguishing between the local states (such as Pará and Paraná) and the Federal State, and in juxtaposing and contrasting their objectives and operations. It should be clear from the above, however, that *both* participate in the political in society, and both are necessarily implied in any theory of the 'State' in this society.

Such a theory emerges slowly through the progressive analysis of the different forms of mediation which are law, bureaucracy and violence, which are finally understood as characteristic of a particular form of the capitalist State – the authoritarian capitalist State. The genesis and formation of this State are seen to be determined by its special relation to the economic in society: it reposes upon and guarantees the reproduction of social relations which are far from being homogeneously capitalist; on the contrary there exists a complex articulation of different modes of production. This heterogeneity at the economic level determines its primary political tasks of controlling labour and underpinning the forms of appropriation and transfer of surplus (often *across* modes of production). The appropriation and transfer of surplus from the frontier is one moment of this general process of economic accumulation. Moreover, this form of the capitalist State does not change over the period of the pioneer frontier; if the continuities at the political level are more important than the changes over this period, it is because 1964 witnesses a change in the form of regime, but not a change in the form of State.

It is impossible to anticipate here the full analysis of this State. All that can be advanced is the notion that this State, given its social bases, is incapable of mediating the rule of the bourgeoisie through mechanisms of consensus and consent (with mediations such as universal suffrage, equality before the law, representative institutions and all the other political attributes of a national 'citizenry'). In fact, the 'incapability' is not in the State, but precisely in the bourgeoisie, which – it is now broadly accepted – is not 'hegemonic'. It is economically dominant but not politically directing in the sense of forming the society in its own image. None of this means that this class must rule uniquely by *force* (although some contemporary appearances in Latin America might lead us to believe so); but it must use different forms of mediation, which both include violence, and which themselves may precipitate or catalyse the exercise of violence.

It is this State which intervenes on the frontier to promote and complete the cycle of accumulation, and through its legal mechanisms and the operations of its bureaucratic agencies acts to mediate the struggle for land. This struggle is nearly always violent, and it is the violence which strikes the attention on first approaching the frontier itself. The violence is integral to the struggle, both mediating it and resulting from it. Other forms of mediation reverberate with the possibility of violence. The classes in struggle live the violence unequally and view it differently one from the other. Violence pervades perceptions and practice on the pioneer frontier. It is with the violence that this study of the frontier begins.

Violence on the frontier

The question of frontier expansion must be posed first and last at the economic level. Finally the process can only be captured within the conceptual framework of the cycle of accumulation which alone is capable of integrating into the analysis all the complex relations of the social reality. First of all, however, it is simply the economic process of the occupation of the land.

Peasants come to the frontier in search of land to settle and so provide for their subsistence. They and their families supply the labour to clear the land, which they claim by their occupation of it. The journey to the frontier may be long and hazardous and the work of clearing arduous. But the peasants have heard the word of the 'common land', the 'free land', the 'land of the nation' (Keller 1973), which they may take for themselves. They press forward in the hope of land to have and to hold. It is their activity on the ground which makes the frontier.

This initial occupation of the land combines abundant labour and land in a spontaneous growth of subsistence agriculture which requires neither infrastructure nor a market. The peasants clear a space in forest or scrub for cultivating the traditional staples (maize, manioc, rice, beans, plantains) or raising a few pigs. Farming is extensive, by slash and burn techniques, with little or no animal traction, and the hoe the only instrument of cultivation. Soon more peasants, perhaps relatives or friends, arrive and claim adjacent plots or buy or receive some of the land already claimed. As occupation intensifies so production increases and the peasants begin not only to produce for subsistence but to negotiate a surplus. Small 'centres' (*centros*) and 'villages' (*povoados*) grow up in such areas for marketing crops and providing basic services. These services include the sale of necessities and luxuries (kerosene, salt, hardware, alcohol, tobacco) and the many bars, hotels and brothels so typical of the frontier town. In short, it seems that occupation of the land will lead to settlement.

But the peasants' hold on the land is precarious and they may not enjoy possession of it for long. This precarity is partly intrinsic to the process of occupation itself, which sees a progressive reduction in the size of peasant plots as its intensity increases. This tendency for the small-holdings to become smaller may combine with a rapid decline in the fertility of the soil to cause crop yields to fall sharply after very few years. The weeds which were destroyed by the fire return and the land no longer 'gives'. This is the easily recognisable malaise of minifundio, which may force the peasant to move forward to the next frontier. But, just as minifundio can only be understood in the context of the near monopoly of land-holding in the Brazilian countryside in general, so the precarity of the pioneer peasants' hold on the land is only comprehensible in terms of the *reproduction* of that monopoly on the frontier.

Peasants claim the land by their labour on it and occupation of it. Their claims are nearly always contested, however, by local land-holders, regional 'political chiefs', or more or less distant entre-preneurs. These large land-holders and big companies assert their 'rights' to the land against the 'claims' of the peasants, and attempt to appropriate the land which the peasants have occupied. Significantly the 'rights' of the economically and politically powerful will very likely not prevent the peasants' occupation of the land, but only facilitate their final eviction from it. In this way a prospective cattle-rancher, for instance, can profit from the peasant labour of clearing the land, by putting down pasture and raising cattle in place of people. In general, it is not only land which is appropriated but the value created by peasant labour in the process of occupation.

This pattern of appropriation is nothing new in the history of the frontier (a fact established in Chapter 4), and it is perpetuated contemporarily on the 'new' frontiers. The roads built from Brasília to Belém and Brasília to Acre, not to mention the Transamazônica, provoked a rush for land among entrepreneurs from Bahia, Espirito Santo, and Goiás, and companies and consortia from São Paulo, Rio Grande do Sul, Paraná, and even the United States of America. The peasants who laid claim to the land find that it is 'bosses' land', 'legal land' like that they left behind, and are forced to leave (or, more rarely, to do 'hired work'). But their pioneering activity was already a result of a monopoly of land elsewhere in the Brazilian countryside, which has today left an estimated six million peasants landless (*Estado de São Paulo* 1975). As land is their means of survival they cannot capitulate so easily in this unequal competition for land. They face the competition by clinging to the land, and the economic process of occupation becomes a political struggle torn by violence.

Academic analysis has not taken sufficient account of this violence. On the one hand, most discussions of violence in the countryside in general have referred to the extra-economic coercion exercised on the large landed estates (Andrade 1963), to the fighting and feuding between local political bosses (Pereira de Queiroz 1969) or to the era of the *cangaçeiros* (Faco 1965); on the other, the best known of the frontier studies have tended to ignore it. Monbeig's study of the coffee frontier traces the expansion of the large estates of the São Paulo entrepreneurs, and where he does encounter frontier settlement by small farmers it is in the highly atypical case of the north of Paraná, which saw the most ordered colonisation ever experienced in Brazil (Monbeig 1952). Jean Roche, in his meticulous studies of the settlement of Rio Grande do Sul and Espirito Santo by German migrants and small farmers (Roche 1959, 1968), favours an analysis of the pattern of economic development and pays scant attention to its political context. Indeed, far from being presented as violent, the frontier has often been viewed as a 'safety-valve' which releases the social tensions in the countryside at large by providing possibilities for movement and improvement and so reducing the prevailing incidence of violence. Finally, where violence on the frontier cannot be ignored it is not explained, but simply classified as criminal (Fontana 1960).

In this respect academic analysis has not advanced beyond the dominant ideological view of the frontier violence, but has rather accepted the ideological categories as a true representation of reality. The purpose of these categories, which themselves resemble the nice distinctions of academic analysis, is to divide the frontier peasants into different *types* of social actor, and then blame the violence on one 'criminal' type. In this way the class nature of the struggle for land is negated, and the violence 'explained' at the level of the 'conspiracy'.

The pioneer peasant is known as the *posseiro*, and the complex range of ideological categories, which vary in time and place, have tended to coalesce around a broad but basic distinction between two types of *posseiro* (Foweraker 1974). On the one hand there is the *posseiro* who occupies the land not to cultivate it but to sell it. He is probably the first to the frontier and works to stake a claim (*posse*) which he can then sell to another peasant. This type is often referred to as the *desbravador de mato*, or 'forest-cutter', and in this interpretation does not remain long on the land, but shifts repeatedly from one claim to the next. The reality underlying the type embraces the many, mostly unrecorded, transactions where claims are indeed sold, not only for money, but for pigs, cows, revolvers, women, and other frontier currency. On the other hand, there is the *posseiro* who does not wish to sell the land but to farm it. He may arrive later to the frontier and buy a claim rather than stake one out for himself. He wants to work the land, has probably paid for it, and so will be reluctant to move.

The second type of *posseiro* is seen as more or less socially 'acceptable'. He works the land and produces. When joined by family and friends from his region of origin he will form frontier communities and begin to civilise the jungle. And, in fact, as many as twenty families migrate together and cluster in the communities of this 'domestic colonisation' (*colonização mansa*), which brings cohesion to frontier society. But the forest-cutter is viewed with more ambivalence. He leads a wild and predatory existence and becomes brutalised by the nature of his work. He is confused, sometimes in fact but more often in fancy, with the criminals who escape society's retribution by living at its edge – on the frontier. Therefore if he is not criminal himself he is infected by an atmosphere of revolt and has nothing but contempt for 'owners'; in his work, possession of the land is its only true title.

Once these types are clearly established it is relatively easy to 'explain' the violence in one of two ways. Firstly, and true to the idea of 'conspiracy', it is possible to represent the 'forest-cutter' as a criminal minority, which will even engage in a '*posse* industry' (perhaps directed by 'subversives'), and invade land already claimed by others, or land which is in dispute, and afterwards sell it again, or demand compensation for withdrawal. Secondly, after recognising that frontier 'farmers' are prepared to respect boundary lines between claims, a more general 'explanation' can focus attention on the 'internal' antagonisms between the two types of *posseiro* with their contrasting behaviour patterns and contrary economic interests. Not surprisingly, academic analysis has taken up the latter, and supposedly 'structural' approach. Velho, in his monograph on Marabá (Velho 1972) speaks of a climate of violence existing between 'more marginal elements' and the farmers; and Monteiro, writing of the migrants of the north of Paraná, finds attitudes either of total conformity or of extreme unrest and revolt (Monteiro 1961). So the typologies which distinguish frontier farmers of 'good faith', on the one hand, who wish to develop a stable pattern of settlement, and the 'marginal minority' on the other, who revolt against the imposition of such a pattern, allow the latter to carry the blame for conflicts over land.

The peasant view of the violence which appears in the popular lore of the frontier is very different. In broad terms peasant perceptions are mystified but reflect their lived experience of subjection and exploitation. The *posseiro* is not seen as actively shaping frontier society, but, on the contrary, as passively moulded by the environment (Westphalen 1968). Just as violence is done to nature, so violence is done to men, and the rules of human interaction match the harshness of the frontier. The peasants do not refer to the misery of a life of bare subsistence, with no vestige of comfort or security, but to the prevailing economic activity on the frontier, which is predatory. In this aggressive atmosphere man degenerates, he attacks nature in search of survival but in the end 'it is the man who is finished'. Moreover, these perceptions are compatible with more general 'explanations' of the poverty and disease of the frontier population as the results of certain characteristics of the 'race' or 'blood' which make the peasant incapable of improvement, and 'uncooperative'. There is no possibility of progress on the frontier because 'here in Brazil everyone wants to eat everyone else'.

Even at this level peasant perceptions are more successful in pinpointing the causes of violence than the dominant ideological categories, if only because personal vendettas and predation of the environment no doubt contribute to the violence. In particular, it is lumbering activity which is notorious in this respect; on every frontier literally hundreds of 'undercover' sawmills whine away during the process of occupation, trees even being cut under cover of darkness by peasants employed by the companies and armed with chainsaws (CODEPAR 1964). Lumber interests often provoke conflict and encourage invasion of land in order to put property rights in doubt (Shigueru 1972; Souza Melo 1976), so as to extract the trees more easily. Finally, however, explorations of this order must refer to the different fractions of capital ('extractive'; 'commercial') which compete in the appropriation of the land, and of the frontier surplus (see Chapter 6).

Moreover, frontier lore also points occasionally to certain regional groups as responsible for the violence (Keller 1973). Such groups may include southerners, *paulistas, mineiros, bahianos, capixabas, goiânos*, depending on the period and place of the frontier. This peasant name-calling successfully identifies as a group not the peasants themselves, who may also have a predominantly regional origin, but the large landowners and entrepreneurs who threaten the peasants' possession of the land. In referring to those 'others' from 'over there', far from their world, the peasants conceive of the antagonism at the level of the secondary, cultural or regional, contradiction but, inevitably, throw into relief the primary class contradiction between themselves and the large owners as the cause of the violence. Their view may be provincial but it is shaped by struggle.

The peasant view goes a long way towards capturing what is common to these 'others', who may appear on the frontier in different guises – or may never appear themselves at all – and so make 'their' identification difficult. The 'others' may include local landholders and politicians, individual entrepreneurs, and large economic enterprises associated with both national and international capital. They assert their 'rights' against the peasants' claims to land, although they may have no better title and far less right to the land than the peasants themselves (who, in law, may assert 'right of possession' by occupying and working the land). They accuse the peasant *posseiros* of invading land which is private property or of criminal practice of a *posse* industry, but among their number must be included the land-grabbers and land-speculators (*grileiros* in the vernacular) who attempt to validate their claims through fraudulent titles to land, which are corruptly issued, or through forged titles which are, of course, never issued at all. In short, while some of the title-holders arriving fresh to the frontier may be genuine applicants for land, many are simply speculators looking to amass large estates and cheat the peasants of the land. Such divergent motives, however, merge in the objective class interest in monopoly of land, which sets the peasants against these 'others'; they are 'others' in the essential sense of seeking profit from land, and not simply survival.

In their struggle for land the peasants confront not only the 'others' but also 'their' representatives both 'public' and 'private' (Ianni 1977). The public representatives are principally the cadres of the bureaucracy and legal apparatuses of the State, while the private representatives are the gunmen hired by the others to protect their land from invasion or to clear peasants from land which they claim. The primary antagonism on the frontier exists between the peasants and the 'others', but, as suggested above, this antagonism is mediated by the operation of law and bureaucracy, and the direct exercise of violence.

The bureaucratic cadres present on the frontier include the officers and employees and technical advisers of local state departments and extension services, and Federal State agencies, especially land and 'development' agencies. The law is manifest in police, lawyers and judges. It is relatively rare for the army itself to intervene on the frontier, except in cases of widespread unrest, but many of the administrators are in fact military men. The gunmen are the 'owners' bandits (*jagunços*); they are usually peasants who try to escape the burdens of class and poverty by joining the oppressors (individual rebelliousness is socially neutral – Hobsbawm 1959). Some may be condemned criminals (recruited from the state prisons, it is rumoured on the frontier); others come from outside Brazil. Private and public representatives may join forces as when gunmen and police operate together, or as when gunmen (*mata-páu*) are employed by the police. But even where they do not, the private practice of violence and intimidation, and the public mediation of the struggle are closely linked.

The peasants on the frontier experience violence not occasionally but persistently. It pervades their struggle for land both in their confrontation with civil society in the form of the 'others' and in their contacts with the State. Like all peasantries, they themselves have no representation at the level of the State, and it becomes clear through their contacts with the State that the range of legal and political apparatuses present on the frontier are finally arrayed against them. This is not to deny the conflicts and contradictions among these apparatuses, nor the limited victories won by the peasants in their struggle, both of which will become apparent in the subsequent analysis; it is merely to assert that (private) violence and State apparatuses complement each other in mediating the struggle for land, and hence the cycle of accumulation on the frontier. In the final part of the chapter this and similar assertions will be made more concrete by summary reference to case material from Paraná and Pará.

Violence and mediation in Paraná and Pará[1]

The direct exercise of violence against the peasantry is usually designed either to extort payment for the land they occupy or to expel them from it. Such violence was used by the colonising company CITLA against the peasants of the west of Paraná in 1957. The company tried to sell its 'titles' to the peasants, who were reluctant to buy what they knew to be worthless documents. Company gunmen then began a campaign of intimidation, to persuade the peasants to pay or to go (and leave the land free for speculatory sales to fresh migrants to the frontier). Houses were burnt down, livestock slaughtered, women and children abused. Peasants were shot down and buried in their own fields. Many peasants agreed to pay, and paid several times over; others fled (and contemporary reports told of at least five hundred families which had crossed over into Argentina). This campaign was finally halted in dramatic fashion by the peasant revolt which swept the region later in the year. But the postscript to the story is written in the events of 1970 in Chopinzinho, part of the area of CITLA operations, and prevents an easy conclusion of a peasant 'victory'. Here – as I learnt in my visit of their winter of that

1. Most of the material for this section was collected from interviews and primary sources. The principal informants are listed, by location, at the end of the bibliography. The written sources are all included in the bibliography itself, but are summarised below for ease of reference:
   a. CITLA and the south-west of Paraná: INDA 1969, Mader 1957, Lacerda 1957, Westphalen 1968, DGTC (Paraná) 1966, GETSOP 1966, *Relatórios* DGTC n.d., *Diario da Tarde, Estado do Paraná* and *Tribuna do Paraná* April 1957 to October 1957.
   b. Chopinzinho – south-west of Paraná: *Relatórios* 1968, 1970 a and b, Tourinho 1968, INDA 1969, Foweraker 1971.
   c. Conceição – south of Pará: Ianni 1977, Souza Melo 1976, DTCC 1974.
   d. São Geraldo – south of Pará: Ianni 1977, Seffer 1976, *O Liberal, Isto é, Mundo*. Portela 1979.

year – gunmen were still forcing peasants to pay for land they had farmed for up to twenty years, and the killing continued. 'Whoever fails to make his peace with the gunmen, dies.'

A similar practice of violence was apparent in Conceição do Araguaia in the south of Pará in the 1970s. There the rush for land heralded by the link road (PA150) between the region and the Belém–Brasília highway brought large owners and small peasants into direct conflict. The *posseiros* were driven from the land by similar threats, and compressed into villages, while the large spaces between were occupied by cattle ranchers. In Rio Maria there were beatings and shootings; in Agua Fria the water course was poisoned. Within the whole area some two hundred murders have been recorded since 1969, but most go unrecorded. As late as January 1977 at least 27 people died in the area of Redenção – as was confirmed by my visit to the graveyard (which grows faster than the town itself). Yet more recently a similar struggle between peasants and gunmen in São Geraldo has made local headlines.

The operation of the law and the legal system complements this violence in at least two ways. The most direct channel is through the collusion and complicity of both agents of the law and of law enforcement in the work of the gunmen. In the case of CITLA the peasants came to recognise that not only were police and gunmen operating together, but that in most municipalities the Chief of Police, Prefect and Judge were in league with the company. In short, the peasants had no recourse to any public figure of justice, and lawyers who tried to defend the peasants were persecuted by the local authorities. Little had changed in this respect by 1970 in Chopinzinho, where a Chief of Police of Pato Branco was again in the pay of the gunmen, and local judges under severe pressure from within and without the administration to find in favour of the gunmen. In Conceição the scenario was set by the killers who walked the streets of the town, so confirming the common knowledge that they were in the pay of the police, and it was again a Chief of Police who directed the corrupt practices of gaming, blackmail, protection and false arrest. Finally, in São Geraldo, the peasants killed two Military Police who were accompanying a land survey which they saw as prelude to expulsion: this desperate act of fearful men drew a rapid reaction in the form of a detachment of two hundred Military Police and upward of one hundred arrests.

More indirectly the law complements the violence through the legal confusion over title which contributes to mask or deny the peasants' right to land, and facilitate their eviction from it. In this connection it is apparent that the usual background to violence includes the tortuous legal debates which characterise the history of the frontier regions. CITLA's 'title' to land emerged from a series of concessions, cancellations and compensations which left the 500,000 ha of the Missões estate in the care of SEIPU (the Federal agency responsible for all 'nationalised' property); the land passed to CITLA by a spectacular piece of legal chicanery, whereby the transaction was registered in Rio de Janeiro by a public notary who was the father-in-law of SEIPU's Superintendent, whose chief assistant was the son of the same notary, and therefore his brother-in-law, and one of the principal shareholders in CITLA. Although this transaction is declared illegal and the case passed to the Federal Congress, CITLA is free to operate on the land until Congress finally makes its decision in 1957.

This confusion and fraud in CITLA's appropriation of the land continued to make property in land provisional rather than definitive in Chopinzinho in 1970. The situation was exacerbated, as elsewhere, by the organised forgery of titles. One group of speculators, in particular, forged documents, elaborated legal 'proofs', corrupted witnesses and bought officers of the law, in a clear effort to gain their ends through the courts. Once land was placed *sub judice* and property rights suspended they could then extort payment for their agreement to drop the suit.

Similarly in São Geraldo titles were issued one on top of the other, causing legal doubts and uncertainties over property rights. The first title was issued by the governor of the state to the Central Brazil Foundation in 1945 and by 1960 a further 490 titles had been issued from this original concession. But in 1961 the state cancelled the concession and the Secretariat of Agriculture titled the area anew, giving some 15,000 ha to IMPAR, a lumber company. On a much smaller scale IMPAR's pretensions in São Geraldo created similar insecurities to those sown by CITLA in the west of Paraná, and, compounded by the prevailing legal confusion, these led to the usual invasion, intimidation and conflict.

In addition to its specific links with the law, the exercise of violence finds support in the bureaucratic institutions and agencies of the State. This is strikingly clear in the case of CITLA whose operations had the direct backing of the local state executive, which used its influence at Federal level to remove obstacles to these operations, by cutting Federal funds to the Federal colony in the area (Colônia Agrícola General Osório). Indeed the state governor intended that the CITLA operations should satisfy the creditors who had financed his electoral campaign. Further, it is evident that the final contention between local state and Federal State was no guarantee of protection for the peasants. On the contrary, it was in expectation of the imminent judgement of the Federal Congress against CITLA that the state executive began to pay these political debts with peasants' lives.

In São Geraldo the survey which precipitated the final confrontation was ordered by INCRA, the Federal land agency, in October of 1976. But a first INCRA survey had already been completed in November of the previous year, which was speciously intended to issue provisional titles to the *posseiros*, but which met with allegations of protecting IMPAR, which itself claimed a large tract of land occupied by the *posseiros*. IMPAR, as suggested above, had begun to use police for its 'clearing operations'. In these circumstances the second survey was thought to have been instigated by lumber interests, although INCRA denied any sinister designs. A final twist to the allegations and counter-allegations was given by the defence lawyer of the imprisoned *posseiros*, who maintained that the INCRA survey looked to expel the peasants from the land and establish it as the private property of no less a figure than Janio Quadros, a former Federal President.

Violence and struggle

The struggle for land can be seen to be unequal, and it is made more so by the organisation of peasant production, which leaves the peasants isolated and open to intimidation. The 'others' who want the land the peasants occupy take advantage of this and bring pressure to bear on individual peasant families on their plots. This was the strategy adopted by a large banking consortium led by Bradesco, for instance, on the 170,000 ha of the Rio Dourado project in Pará: one by one the peasants who had farmed the land for decades were 'convinced' to accept indemnities which would not even pay the cost of their removal. In this way the 'others' never confront the peasantry as a class, and class organisations are either not permitted, or where permitted, closely overseen by the State and manipulated by the 'others'. The Union of Rural Workers of Conceição, for example, faced threats of death on its inauguration in 1971 and the Ministry of the Interior delayed its official recognition for some three years, effectively haltering its activities over this period. Following its recognition in 1974 it was immediately reviewed by the Ministry of Labour, which dismissed its secretary and lawyer.

In these circumstances, it is the individual peasant who struggles for land. His individual reaction to violence will almost certainly lead to defeat, even death. Moreover, insofar as the continual practice of violence is used against individuals, it is largely unseen and his resistance will go unrecorded. Exceptionally, however, the peasants succeeded in organising resistance to the violence, and it is this organised resistance which most clearly reveals the practice of violence on the frontier. On these occasions the peasants cease to be simple victims of the violence, and become political actors in the class struggle.

In the west of Paraná in 1957 the growing opposition to the company gunmen was first evident in the petition with some 2000 signatures sent to the Federal President in June, and in the following months the gunmen increasingly met with armed resistance. But the signal for revolt was finally given by the Federal Congress which denied CITLA's right to the land, and the following day, 11 October, the peasants of Pato Branco and Francisco Beltrão rose en masse, and set up a General Assembly of the People to direct resistance to the company. Other municipalities followed suit and organised resistance throughout the region. Local state authorities then had to treat with the rebels, and effectively acceded to their demands, which included closure of company offices, dismissal of lawyers and judges who had connived with the company, and recognition of the peasants as the rightful owners of the land.

The case of Paraná is exceptional even among these exceptions, but wherever peasants succeed in organising in sufficient numbers they have some chance of winning at least a local victory. On the Floresta estate in Pará a bishop assisted the peasants in settling the public land of the area, but, predictably, it emerged that at least half of these lands already had title. However by that time there were upwards of 2000 peasants on the estate and it appeared likely that the State would recognise the existence of a 'social problem' and expropriate the private 'owners'. In short, the State will intervene not to protect individual peasants, but to prevent class revolt. Thus a 'social problem' is defined not merely by the absolute number of peasants involved but by their concentration in one area and consequent potential for revolt.

Another exceptional case was that of the Araguaia region in the south of Pará in the early seventies, which harboured a rural guerrilla movement in the area of São Geraldo, Perdidos and Boa Vista. The peasants did not initiate this organisation, but the movement found its social base in the small settlements of *posseiros*. It would be easy to overestimate the significance of the experience, and it appears that at its peak no more than fifty to seventy persons were under arms, acting in three groups of twenty or so. The guerrilla operations continued over the three years before 1974 and were finally halted by massive intervention by the three branches of the armed forces. Probably it was the peasants, once again, who bore the brunt of the military onslaught. Ironically, while there are no other known cases of organised military resistance on the frontier over the last decade, wherever peasants begin to band together in their own defence, however legitimate, they are immediately accused of infringing the laws of national security.

Thus it is only through organising as a class that the peasants can gain any protection from the violence. They have no recourse to the law, which is beyond their reach and beyond their resources and which anyway acts against them. They have no recourse to the bureaucratic agencies of the State, which themselves have no capacity and no policy for protecting the peasants. In fact, their only consistent ally in the contemporary context is the Church – or sectors of it. The three 'red bishops' of Marabá, Conceição and São Felix do Araguaia (in Amazônia) speak out on behalf of the peasants and are accused of 'stirring them to revolt'. Two of them attempted to intervene in the São Geraldo dispute by denouncing the violence and the injustices. The State replied by imprisoning their emissaries to the military command in Marabá, and by publishing false reports of their interrogation. This manoeuvre appeared designed to reduce the impact of a radical pronouncement made by the Church at that time on the plight of the peasantry ('Pastoral Communication to the People of God' CNBB 1976). The two bishops themselves were interrogated in Belém with the intention of inculpating the Church in the conflict and condemning it too of 'subversive activity'. The National Council of Bishops of Brazil and the Land Pastoral (Pastoral da Terra) constitute the last line of defence of the dispossessed.

Conclusion

This summary account of the violence on the frontier indicates its close links with the operations of the legal and administrative apparatuses of the State. It therefore makes no sense to insist on *private* corruption and violence which exist *despite* the intervention of the State on the frontier. In the case of the law, although criminal activity from forgery to murder abounds, and lawyers and judges are witting or unwitting accomplices in crime, the violence cannot be explained as a result of 'lawlessness' on the frontier – if only because legal confusion invites the violence and legal decisions often unleash it. Similarly in the case of the administrative and repressive apparatuses, the violence cannot be a result of their 'ineffectiveness' – if only because it is often exercised by the police, and more rarely by the army, or may itself result from administrative initiatives. In other words, interpretations which would explain the violence in terms of the 'lawlessness' and 'administrative ineffectiveness' on the frontier, as opposed to the rule of law and State control in *national* society, point to the *means* of violence rather than its cause. Violence, law and bureaucracy work in complementary fashion to mediate the struggle for land on the frontier.

The violence is not therefore explicable at the political, but only at the economic level. Simply put, the violence, 'lawlessness' and State 'ineffectiveness' are all parts of the complex economic process of the occupation of the land by the peasants, and, crucially, their final expulsion from it. The process provokes administrative conflict, legal conflict and finally physical conflict and violence. Widespread violence is in fact the expression of a struggle for land, and, in broader terms, a struggle for the value created in this process of economic expansion. It is a class struggle waged over the appropriation of this value, and over the peasants' right to survival. Although divergent interests emerge between different landowners and economic enterprises, between different fractions of capital, the primary economic and political antagonism lies between these dominant classes and the peasantry. The violence against the peasantry and the roles of law and bureaucracy in this process are economically *possible* because prevailing economic conditions of a large labour surplus and a constant flow of fresh peasants to the frontier relieve the dominant classes of the necessity of guaranteeing the reproduction of the labour force on the frontier (this very process of frontier expansion guarantees this for them); they are economically *necessary* to win the struggle for land and by expelling the peasants or pushing them into areas of minifundio to achieve and reproduce the cycle of accumulation on the frontier, which simultaneously reproduces the conditions of accumulation in the Brazilian countryside, founded as they are in the monopoly of land.

The following chapters will investigate the different dimensions of frontier expansion and the struggle for land. As suggested above, the analysis of law and bureaucracy in the mediation of the struggle (Chapters 4, 5, 6, 7) will in general precede the integrated account of frontier expansion at the level of accumulation on the frontier (Chapter 8) and accumulation in the national economy (Chapter 9), and finally at the level of national political economy and the State (Chapter 10). Nonetheless not all economic analysis is left until the closing chapters. The operations of the bureaucracy are of interest precisely insofar as they interact with and 'translate' private economic interests: therefore the role of private capitalist enterprise, especially colonisation companies, in their relation to the State land agencies is discussed in Chapter 6; while Chapter 7 examines the changes in the relationship between capital and State on the pioneer frontier over the period of its recent history. More immediately the study proceeds to carry out the economic groundwork for later analysis by establishing the reciprocal relationship between frontier and national economy in Chapter 3, concentrating on the process of industrialisation and the changing markets for land, labour and goods; and by exploring the process of frontier expansion itself in terms of the changing production and market relations on the frontier and of its progressive integration into the national economy, which is material for the next chapter.

2

The process and stages of occupation of land on the frontier

The process of expansion of the pioneer frontier describes the progressive integration of the frontier region into the national economy. The same process contains the cycle of accumulation on the pioneer frontier. In other words, the cycle of accumulation runs its course through the integration of the frontier region into the national economy. This national economy is capitalist, where accumulation takes place through the appropriation of surplus value; the frontier economy is not originally capitalist, but, on the contrary, is characterised by clearly 'pre-capitalist' production, and occasionally market, relations. Thus the transformation of the 'natural environment' of a frontier region into a 'productive society' describes the *transition* from pre-capitalist to capitalist relations occurring within the cycle. Given the heterogeneous structure of the Brazilian social formation, where different modes of production exist side by side, the transition implied by the cycle may never be complete.

The concept of transition implies changes both in production relations and in the markets for goods, land and labour. These changes are complex, and take place over time. But it is possible to capture the principal dimensions of change by distinguishing three consecutive 'stages' of frontier expansion, called here the 'non-capitalist', the 'pre-capitalist', and the 'capitalist'. There are doubts which quite rightly surround any division of social process into 'stages', and the strategy is adopted here as a heuristic device which should not be taken to deny the central idea of the *process* of frontier expansion. These stages should not be viewed as discrete and mutually exclusive categories in the manner of 'ideal types', but merely as moments in the process of transition; such that elements of the first stage may subsist in the final stage, while features of the final stage may be found at the beginning of the process. Transition from one stage to another occurs at that moment in an increasing social division of labour when new productive relations emerge as dominant in the region. At the same time, however, the unimaginative but accurate labelling of the stages betrays their teleological character. The stages make sense insofar as they order the description of the transition to the dominance of capitalist social relations.

The immediate effect of the 'stages' is to demonstrate that the diversity of frontier experiences is in part only apparent: frontiers may appear different at one point in time because each frontier is different from itself at another time. Moreover, the further demonstration that similar stages on different frontiers are comparable in the composition of their production and market relations – despite the different periods of their expansion – provides support for the interpretation of all frontiers in terms of the cycle of accumulation which is essentially the same on every frontier. Frontiers in the west of Paraná and south of Pará expand over different periods, but the division of each 'process of integration' into the same stages is not difficult to sustain. Finally the division of the process into stages creates a conceptual framework which can inform the investigation of the various empirical variables evident in descriptions of frontier expansion, such as population growth, access to the region, volume of production, integration into the market, and prices for land. In this investigation the question of migration to the frontier will be considered an 'independent' variable, which will not be explained until the following chapter.

The non-capitalist stage: Paraná and Pará

At the beginning of the process of expansion the frontier economy is extremely isolated, and its activities largely extractive. The sphere of exchange is limited, depending on barter internally, and on outside commercial centres for placing one or two products in a regional market. This precarious market for goods marks the initial stage of the integration of the region, but there is as yet no market for either land or labour. The social relations of production in the region are mainly servile, and manifest all the signs of a direct coercion of labour. But there also exists an emerging petty commodity sector, and the significant historical result of this stage is the incipient peasant class which begins to occupy the region.

The west of Paraná was unpopulated until the end of the nineteenth century (Plaisant 1908), and after this its few inhabitants were employed by the companies which had received land grants from the local state. These companies sought to exploit the maté, and later the lumber reserves of the region, and used their private police to prevent settlement as such, and to control their landless and nomadic workforce (Correa 1970). To the north, the Maté Laranjeiras company, in similar fashion, was using its economic power to manipulate the state of Mato Grosso, and its police power to expel *posseiros* from the land it controlled (Correa Filho 1957). To the south the sufferings and insecurity of this workforce was exacerbated by the rivalry of the state governments of Paraná and Santa Catarina, which attempted to assert their dominance over the contested areas of their western border by land grants to local politicians. The oppressive and cruel conditions experienced by the first frontier peasants goes far to explain the fanaticism of the War of the Contestado (which was, in fact, a prolonged revolt which continued throughout the late 1910s and early 1920s) (Pereira de Queiroz 1957; Vinhas de Queiroz 1966); these conditions were documented for the first time by João Cabanas and Juarez Távora who marched through the region in 1924 at the head of the Revolutionary Column of Luis Carlos Prestes.

During this early history of the region settlement was slow, sparse and scattered. Where the companies were present they actively discouraged settlement; where they were absent some few inhabitants who had filtered in along forest trails lived in small *caboclo* communities. In general, the lack of settlement was simply due to the region's isolation (Correa 1970). Sporadic attempts to colonise had met with small success for the same reason: Federal efforts to build 'military colonies' along the border with Argentina had failed; the state had granted land to colonising companies which conspicuously failed to colonise. As late as 1943 the governor of the Federal Territory of Iguaçu (one of the initiatives of Vargas' 'March to the West' which included the west of Paraná) complained that he could not reach the region for lack of access: from São Paulo the journey took at least seven days by boat and train; the trip from Curitiba, the state capital, was considerably longer.

In these conditions it is not surprising that economic activity in the region remained relatively undeveloped. It is true that despite the isolation economic changes did occur with Argentine self-sufficiency in maté and the decline of the maté economy in the 1930s (Carneiro n.d.). In the following years the first migrants from the south began to arrive in the region and within a short time pig-rearing emerged as the principal economic activity (Correa 1970). This is significant in that it required the peasants to clear the forest and not merely to penetrate it: even if pigs were first reared loose in the forest, they had then to be sent to the *safrista* for fattening, and he at least was obliged to clear ten or twenty hectares for the planting of maize. But, in market terms, little changed: on the one hand barter continued, and pigs as much as maté could be exchanged for goods such as salt, cloth, iron-ware, alcohol and kerosene; on the other dependence on outside commercial centres such as União da Vitória continued. Pig-rearing was only possible in these conditions because a pig is its own means of transport.

Pig-rearing was an extensive activity, which was visibly incompatible with an intensive occupation of the land. Reared loose in the forest each pig required five ha; the ratio at the *safra* was five pigs per ha. But with land abundant and occupation only beginning the peasants could appropriate sufficient land for their subsistence needs. At this time the land was freely available: no market existed for it and therefore it had no price. The peasants produced primarily for their own use, and while some goods did enter a market, the economic activity of the region was far from being structured by that market.

Coincidentally the occupation of the south of Pará also began at the turn of the century, when the rubber boom arrived in Conceição do Araguaia. Prior to that time the only occupation the region had known was by Indian tribes, and one or two isolated fazendas. Like maté, rubber collection is an extractive activity, and, in the case of Conceição, like maté it is nomadic. This parallel with early economic activity in the west of Paraná derives from the technology of collecting *caúcho* (as opposed to *borracha*). With *caúcho* the rubber tree must be felled before the latex is extracted, implying a constantly moving work-force. It also implies similar problems of labour control, and a similar exercise of extra-economic coercion. With the arrival of the boom the hired gunmen of the rubber boss drafted the Indians to work as rubber collectors and carriers, and secured de facto appropriation of the rubber areas (Ianni 1977). More labour was imported to the region and both Indians and migrants suffered a similar oppression to the *caboclos* of Paraná. Moreover, in the case of Pará, the work-force came to be controlled through the notorious system of *aviamento*.

The rubber collectors lived within the forest (in the 'centres') while the local rubber bosses or *seringalistas* lived on its edge (in the 'margins'). They were generally paid not in money, but in food, tools, clothes and medicine (and alcohol, tobacco and women). In fact, the necessities for their work and survival were 'advanced' to them, including the price of transport to the region, so that they incurred debts from the beginning of their contract. These had then to be paid by their production. But by raising the price of these necessities, and by other ploys, the rubber boss made sure the debts were never paid and the collectors never free of their obligations. Moreover, not only were the collectors not allowed to plant for their own subsistence, but the boss also exercised a monopsony: to sell to another boss was 'stealing' and payable with death. This a priori and a posteriori system of indebtedness, articulated through the *barracão*, the boss's shop, operated to control labour in the rubber economy (Ianni 1977).

Only a highly profitable product like rubber during the boom could have brought the south of Pará within the ambit of the national economy in that period. The region itself remained more isolated than the west of Paraná, and until 1940 the journey from Conceição to Belém, the state capital, along the Araguaia–Tocantins river took at least forty days. As opposed to what occurred in Paraná, Conceição did develop as a commercial centre in its own right, but direct dependence on the outside commercial centre continued. This was largely due to the structure of indebtedness of the *aviamento* system: just as the rubber collector was in debt to the rubber boss so he in turn was in debt to the wholesaler and merchant in Belém (Santos 1968) (and the structure of the system goes far to explain the precocious growth of cities in Amazônia) (BASA 1967). The exporting houses and banks were themselves linked with industrial capital insofar as the rubber was used to produce industrial goods in the industrialised countries, and the 'rubber economy' itself consumed a range of these goods in its activities. In this way even during its early history this frontier economy was subject to a process of industrial accumulation elsewhere.

But the boom died. From producing 100% of world production in 1878, Amazônia was reduced to 12% by 1919 and 2% by 1929. In the south of Pará this economy stagnated and the population decreased. But some of the collectors turned to subsistence farming for survival, and the years after 1920 saw the emergence of a peasantry (Ianni 1977), grouped in small communities, which lived by rudimentary agriculture, fishing and hunting, and artisanal industry. While this was primarily a subsistence agriculture, an occasional surplus was marketed in the town of Conceição, and extractive activity continued intermittently. In the first years this would still have been of rubber (or of the so-called *drogas do sertão*), but in more recent times cashew nuts and certain woods like mahogany have become more important. Very often labour moved between extractive activity, sometimes of more than one product, to subsistence farming, and back again – perhaps on a seasonal basis. In this, its movement mirrored the labour process in Amazônia as a whole (Magno de Carvalho 1976).

Thus these two frontier regions show significant similarities in their first stages of expansion (as does the south of Mato Grosso, as will appear in Chapter 4). In both cases the initial economic activity was extractive, but in neither case did this experience have any immediate or decisive impact on the settlement of the region. This was due in part to their relative isolation and in part to the nomadic nature of the extractive labour process. Yet this initial activity in its decline created economic conditions which would influence the later expansion of the frontier. The most important of these was the emergence of a peasantry which remained in possession of the land, which was abundant, and practised an extensive agriculture. It produced mainly for subsistence, but marketed an occasional surplus. These peasants then remained 'available' in these regions, and in the second stage entered new extractive activities such as cashew nuts in Pará and lumber in Paraná. They also began the process of occupation of land by small farmers, and, in the final stage, some have succeeded in producing commercial crops for the regional and national markets.

In the literature this stage of frontier expansion is often referred to as the 'expanding frontier' where the 'demographic' frontier moves ahead, while the 'economic' frontier lags behind. These categories are misleading in that they imply an absolute lack of economic activity, and they were rejected by Martins (1971) who preferred the concept of a 'surplus economy', where production is primarily for subsistence, but where a 'surplus' may be marketed – using those factors, principally labour, which exceed subsistence requirements. In short, production is of use-values, some of which acquire exchange-value, not as a result of a complex social division of labour within the region but simply because a market exists for them outside the region. This represents an advance in the analysis as it correctly characterises the historical result of this stage which is undifferentiated petty commodity production. But as the productive process during this stage includes both servile and petty commodity relations, here it is simply called 'non-capitalist'.

The pre-capitalist stage: Paraná and Pará

The second stage of frontier expansion is characterised by increased migration into the region and more intensive extractive activity. Land begins to be bought and sold, but the price represents the value of what is on the land, and is not at first the price of land itself. Indeed, the conditions of extended extractive activity can enter into contradiction with an emerging regime of private property in land, and, for this reason among others, extractive activity in this stage tends to have a decisive impact on the pattern of settlement. (There may be attempts to resolve this contradiction, such as the recurrent leasing of land by the state of Pará, for the purposes of extractive activity.) These changes bring about regular production of commodities, and an emerging market for land; but there is not as yet any 'free labour market'. The social relations of production are those of the petty commodity sector, or 'mixed' forms of servile and capitalist relations. Servile relations are residual or reproduced anew; the payment of wages is not always accompanied by free movement of labour and does not always exclude forms of extra-economic coercion.

The rhythm of migration into the west of Paraná rose rapidly in the 1940s (Bernardes, N. 1950), but the influx of migrants did not suddenly transform the frontier into a stable agricultural community. In 1950 the land of the region was still covered with forest, and for some years to come would itself hold little 'value'. Throughout the region it was the trees that commanded high prices, and particularly the rich reserves of pine-wood (araucaria). Indeed, the majority of the so-called colonising companies operating in the region at this time were, in fact, lumber companies in all but name, and the assertion of claims to land was motivated by the pines it supported (CODEPAR 1964). Land was merely a residual investment, and the rising price of pine wood in the national and international market in subsequent years assured the predominance of lumbering as the principal economic activity, overshadowing incipient agricultural growth.

A measure of this predominance is given by the available figures (CODEPAR 1964) which show 80% of all persons employed in 'industry' in the region in the 1950s to be employed in the lumber sector, and 60% of all non-agricultural enterprise in the same sector by the end of the decade. Yet the figures for numbers employed in the industry seem low; only 3900 in 1950, rising to 12,100 by 1960. In fact, the figures fail to capture the absolute scope of the lumbering activity, because the companies were far outnumbered by the small 'undercover' mills and logging camps which accompanied the advance of the frontier. These were manned by labour gangs hired from the available peasantry (*na empreitada*), and controlled by the gang boss. These 'undercover' mills worked the reserves of small *posseiros* and flourished wherever land was in dispute, and the peasants could be driven from it. The peasants were anyway eager to sell their trees at low prices in the knowledge that they could not defend themselves against hired gunmen, or the land against invasion. As the threshold of investment for entry into logging is low, an unknown but very high number of mills operated unhampered (and clearly outside the control of the National Pine Institute – INP).

The big companies also brought pressure to bear on the pioneer peasants as competition between them forced prices down and the pace of production up. As the overall demand for pines continued high they preferred to buy 'outside' their own areas, and forced prices to the small-holder down by fomenting legal disputes; their own reserves were maintained against future price rises and defended by their private police forces. In attempting to corner supplies in this way, company operations might combine with the undercover mills, as when the mills colluded in by-passing INP quotas by sawing logs for the companies in sub-contract. In the same attempt, the companies' 'contract' with the peasants prevented the use of the land until the companies chose to fell the trees – and the peasants suffered not only from 'lost production' but from the inflationary corrosion of the prices they were paid at the time of contract. Thus peasant economic enterprise was checked by lumbering activity (Bernardes, L. 1953), and their possession of the land they farmed threatened by the trees which stood on it.

In 1947 there were 128 million pines in Paraná, 27 million in Santa Catarina and 6 million in Rio Grande do Sul (CODEPAR 1964). The first five years of the fifties saw the complete exhaustion of the pines of Rio Grande, and the progressive exhaustion of the reserves in zones near to Curitiba. As prices were initially a function of transport costs to the nearest commercial centre they began to rise rapidly and companies and mills moved equally rapidly into the west of Paraná. As is apparent in the table below, prices which had merely kept pace with inflation until 1955, accelerated over the years until 1960 and outpaced the overall rise in prices. After 1960 prices rose yet faster as they were now a function of the incipient exhaustion of all reserves (i.e. a function of the expected rise in prices).

Table 2.1 *Prices for pines and pinewood in relation to the general rate of inflation 1949–64*

	Standing pines A[a]	Sawn logs B[b]	Relation A/B%	General index[c]
1949	10	12	83	14
1950	11	15	73	16
1952	13	18	72	21
1955	26	70	37	36
1958	30	80	37	56
1960	*100*	*100*	*100*	*100*
1961	155	160	96	137
1962	450	350	128	208
1963	620	380	163	360
1964	850	400	212	614

[a] SPL – Servicos de Planejamento (from firms records)
[b] Anuario Florestal
[c] Conjuntura Economica: June 1964. Quoted from CODEPAR (1964).

It is therefore not surprising that the rhythm of extractive activity quickened so rapidly during the fifties. At the same time, the finally frantic pace of forest exploitation was matched by a huge influx of migrants into the region, who claimed the land and began to work it.

What had been a sparse and scattered settlement became a populous pioneering movement. It was soon apparent that the predatory and speculatory extractive industry could not but prejudice the equilibrium of the agricultural year. While agriculture required a stable pattern of settlement, lumbering did not; indeed the progressive settlement of the land severely constrained the operations of the lumber companies, precisely during the period of fast rising prices (Bernardes, L. 1953). So they fought yet more fiercely by legal manoeuvres and political pressures to get effective control of the land – which finally paralysed attempts to control its settlement. The peasants, for their part, competed hard to stake a claim and stay in possession of what was very fertile soil. In this way the two economies, extractive and agricultural, moved incongruously yet inexorably closer together both geographically and socially into what proved to be a violent juxtaposition. The struggle of the peasants to stay on the land and produce, in the face of the violent search of the lumber companies for quick profits, determined the direction of frontier expansion in the second stage.

In the south of Pará the potential contradiction between extractive activity and private property in land was postponed for decades by the traditional practice of leasing land, and the late arrival of intense peasant migration into the region. In the *aviamento* system the land had always been leased to the rubber boss, and the practice continued with the cashew nut groves, once cashews found a firm hold in the world market about 1930. The leasing served two purposes. In the first place, the terms and area of lease could be regularly revised to meet the changing requirements of the extractive economy (private property in land could not provide the same flexibility); in the second place it provided an income for the state, which it could not have realised through the sale of land. The economics of extraction did not favour the purchase of large tracts of the land itself, which was not viewed as 'valuable'; but leaseholding allowed exploitation of what was on the land – at a price. Leaseholding continued unperturbed by economic and political changes until 1966.

At the same time the lumber industry was slow to develop in the state, and once a few small mills had been mounted around Belém it tended to remain stagnant. The big market for wood was in the south, but the south had its own sources of supply for the time being – precisely in the pines of Paraná (and rosewood of Espirito Santo and Bahia). In 1945 some small stimulus was given by production for the US market and foreign capital (US, Dutch, Danish, Japanese) began to penetrate production. But the foreigners did not install their own plant; they lent capital to local entrepreneurs who could easily supply the limited demand for certain species of hardwood. Brazilian capital continued to be invested in the south, and production in Pará was confined to small mills situated along the river banks.

The situation was rapidly transformed in the 1960s with the building of the Belém–Brasília highway, and the programme of fiscal incentives to private enterprise in Amazônia. At the same time that forest, and particularly, pine reserves were exhausted in Paraná, the Federal State, through its road construction and through its agency SUDAM (Superintendency of Amazônia) created the impetus for the rapid extension of the frontier in the south of Pará, and elsewhere, through extensive cattle-raising projects. Between 1968 and 1974 SUDAM approved 300 cattle projects in areas of hardwood (Muller and Brandão Lopez 1975), which was destroyed in order to put down pasture. By 1974 the 89 projects in the south of Pará alone had cleared 400,000 ha of forest (Pinto 1976a). Each extension of the incentives – such as the Amazon Investment Fund – served to raise the rate of exploitation and the progressive tax on unproductive land (*Impôsto Territorial Rural*) levied by the State land agency INCRA stimulated the removal of forest cover which simulated investment in cattle-raising. Such was the rapidity of the process that only a fraction of the wood (an estimated 570,000 m^3 out of 4,150,000 m^3 by 1974) was economically exploited rather than simply destroyed (Pinto 1976b). It was in these conditions that the second stage of frontier expansion in the south of Pará ran its course.

SUDAM incentives brought more national capital into the lumber industry in Pará, and, significantly, the new medium capacity mills were constructed by entrepreneurs from Santa Catarina and Paraná. As in the west of Paraná, however, most mills in this second stage remained small and 'undercover'. They proliferated not as a result of SUDAM financing, which they rarely enjoyed, but of increased access to the wood, offered by the new highways (Belém–Brasília and PA-150 Marabá–Redenção). These small mills led the boom, operating out of sight within the SUDAM cattle projects, or on the edge of the frontier. They tended to extract only a few highly profitable species, and so destroyed forest at a high rate. They sprang up in their hundreds along the Belém–Brasília road in the 1960s, and consequently along the 1500 kilometres of the highway the only Amazon rain-forest to be seen is in the Parque Rodriquez Alvez on the road into Belém (Moura Castro 1975).

Large mills and little mills alike sawed wood extracted from the clearing of forest on the large estates. The work of clearing (*derrubada*) was done by labour gangs contracted locally, or in the northeast of the country. They were brought in by truck, or, on occasion, flown in by plane, and although by their contracts they were wage-labourers, they were subject to extra-economic coercion. In some cases they arrived owing their passage to the company, and so worked to redeem the debt as in the system of *aviamento*. In other cases their very isolation and the company police prevented them from leaving the work front. In the event of resistance, at the very least wages were withheld, or perhaps not paid at all.

As for the peasants of the region, they could live more or less comfortably with extractive activity while this was confined to the leasing of cashew groves. Indeed, this extraction, unlike lumber in the west of Paraná, occasionally served as a complementary activity to their subsistence farming. But when with the imminent exhaustion of reserves in the south, the coming of the roads and the initial SUDAM incentives, the extraction moved into lumber, their livelihood was threatened in similar fashion to that of the Paraná peasant. In effect, given the relatively late influx of pioneer peasants into the region, followed closely by the Federal incentives which provoked a rush for land, the second stage of expansion in the south of Pará was considerably 'compressed' in comparison with the west of Paraná. Direct confrontation between extractive enterprise and peasants did not occur until the sixties, when the peasant struggle for land began; by the seventies the frontier was already moving into its third stage, which sees the intensification of this struggle. The length of the second stage is not what matters, of course; only the fact of the transition from non-capitalist to predominantly capitalist production.

The capitalist stage: Paraná, Mato Grosso and Pará

The final stage of frontier expansion is characterised by an intense migratory flow into the region and established access to the national economy. Economic activity is no longer predominantly extractive but agricultural, and agricultural production itself tends to become increasingly capitalised. Concurrently, there is a rapid increase in the price of land, and the pattern of land tenure within the region becomes increasingly concentrated. These developments express a regular and diversified market for goods and a consolidated land market. While petty commodity relations certainly persist, capitalist relations of production are now dominant, as is demonstrated by the growing 'free' labour market. However, much of this labour is still employed on a seasonal or 'task' basis, and the wage relation itself is often weighted with 'pre-capitalist' restrictions. This is one indication that the expansion of capitalism in the countryside reproduces not only the wage relation but also pre-capitalist relations, and hence achieves the reproduction of an articulation of different modes of production.

As the frontier moves into this final stage, economic activity in general is more differentiated than previously, and the social division of labour more complex. At the same time, as the frontier is further integrated into the national economy more information on these economic developments becomes available, although that information is still fragmentary. Thus the intention now is to present a slightly more detailed empirical examination of the different aspects of these developments than was possible for the previous two stages – before attempting any kind of synthesis. Insofar as the empirical material is also 'retrospective' this presentation will also illustrate the complexity of the *transition* occurring within the process of frontier expansion.

Population increase on the frontier

In general, the very intensity of the process of occupation of land tends to distinguish this stage from the previous two. Turning first to the west of Paraná, the figures on population growth, and the approximate figures on population density (table 2.2) demonstrate that the region grew fastest in the decade 1960–70. The figures have been divided into those for the southwest (below the Iguaçu river) and those for the west (above it) in order to demonstrate that this decade corresponds precisely to the third stage of expansion. The frontier advanced from the south, into the southwest, in the 1940s, and although the region felt the full force of the advance in the early

Table 2.2 *Population density and population growth in the west of Paraná 1940–70*

a. Population density (inhabitants/km²) 1940–70

	1940	1950	1960	1970
Southwest	0–5	5–10	10–25	25–50
West	0–5	0–5	0–5	25–50

b. Population growth 1960–70

	1960	1970	1970 Urban	1970 Rural
Southwest	228,923	450,338	81,922	368,416
West	135,677	756,900	151,887	605,013

IBGE: Census data

1950s, the demographic effects were only evident some few years later. Thus there was a pause while the region 'absorbed' the impact, that is, while the 'pre-capitalist' stage ran its course.

In the case of Mato Grosso, the overall population density in the state was so low that the huge migratory flow into the south radically altered the demographic composition of the entire state, as is indicated in table 2.3.

Table 2.3 *Population density and population growth in Mato Grosso 1940–70*

	Inhabitants/km²	Increase in pop.	% Increase	Average rate (per 100)
1940–50	0.42	89,779	20.77	1.94
Pop. 1950 = 516,514				
1950–60	0.74	388,218	74.36	5.62
Pop. 1960 = 910,200				
1960–70	1.30	713,306	78.37	5.96
Pop. 1970 = 1,623,618				

IBGE: Census data

The area of the state was huge (1,231,500 km², or 14.6% of the total land area of Brazil) and its population predominantly rural (60% in 1960 with 70% of the labour force employed in the primary sector). But this population was relatively concentrated, and this concentration reflected the process of occupation of land on the frontier (Alvez de Souza 1965). The population of Campo Grande (96,602 km²), for instance, in the south of the state increased at a rate of 8.1% per year by a total of 164,255 persons during the fifties (which was 42% of the total increase of the state's population over the decade, giving it a population of 303,970 in 1960, or 33.4% of the state total). This massive increase, in turn, was principally due to Dourados, the area of the Federal colony, and the principal area of crop cultivation. Its population grew by 611% over the decade, making it the fastest growing municipality in the state. This concentration of population continued in the following decade, such that by 1970 of the ten most populous municipalities in the state, eight are found within this frontier region. These eight supported 782,283 inhabitants in this year, who composed 45.1% of the state total. Dourados and environs probably contained half of this population.

In the Amazon region as a whole the rural population had grown very slowly from 1950 to 1960 (28%) and the large proportion of the population born within the region in 1960 (93.25%) showed that little of this increase was due to immigration (Magno de Carvalho 1976). Owing to a continuing exodus to the cities, growth was yet slower in the following decade (22%) with two major exceptions. The first was Rondônia, where the same migratory flow that had pushed forward the frontier in the west of Paraná and south of Mato Grosso entered the region; the second was the south of Pará. The population of Conceição do Araguaia (28,572 km^2) grew from 11,283 in 1960, which was roughly equivalent to what it had been in 1920, to 28,973 in 1970 (or 38,038 including Santana do Araguaia, a district of Conceição which had gained municipal status in 1961) and to an estimated 70,179 in 1977 (Magno de Carvalho 1976). As in the case of Mato Grosso, this increase changed the demographic profile of the state by concentrating population in the south, and again this concentration was a result of the arrival of the frontier. The first sure sign of the advancing frontier is the fact that of the inter-state migratory flows into Pará, the urban–rural flow is more expressive than the rural–urban, and so, in 1970, of the economically active population of Conceição 84.4% were employed in the rural economy including extraction (as opposed to 60% in the Amazon region as a whole). This contribution of migration to present population levels in the south is demonstrated in the table below, which specifically indicates the proportion of migrants arriving from outside the Amazon region to Conceição and its neighbouring municipalities.

Table 2.4 *Migration into Conceição and neighbouring municipalities 1960–70*

	Population 1970	No. migrants over decade	% migrants from outside Amazon region
Conceição	28,953	19,255	93.8
Marabá	24,474	11,540	85.4
Santana	9,035	4,601	90.5
São João	15,326	8,562	90.4

From Magno de Carvalho 1976

Access to the national economy

There is a saying in Brazil that 'progress is roads' and in the popular view roads sometimes seem endowed with almost mystical developmental powers. In this account of the pioneer frontier, on the contrary, the road is not seen as having any independent capacity to determine frontier expansion. The immediate determinants of the process must be related on the one hand to the constantly increasing demand for rice, beans, beef and other staples in the major cities of the country and, on the other, to the pioneering movement itself and the forces which drive it (which is work for the next chapter). But the road is clearly a catalyst in the transition from non-capitalist to capitalist production and contributes to remove the frontiers from their relative isolation. The rapid extension of the road network over the central period of the pioneer frontier is therefore important to the analysis.

This network almost doubled between 1956 and 1965 from 460,000 km to 803,068 km (70% of this network still being in the south of the country). At the same time the number of trucks and vans in the country grew at a rate of 12% per year from 1956 to 1960 and 7% per year from 1961 to 1965; by 1966 there were 785,000 vehicles overall and 70% of goods transport was by road (Ellis 1969). The impact on the frontiers under study was instantaneous: in the south of Mato Grosso, any commercial agricultural production had had to be within reach of the North-West Railroad – with the exception of cattle which could walk to market – but with the arrival of the road in 1955 Dourados suddenly 'took off' economically; in the west of Paraná goods traditionally went down the Paraná River to Argentina, and as late as the 1960s plans for improving the marketing network were debating new feeder roads to ports on this river, when the region was surprised by the asphalted BR-227 which linked it directly to the state capital, and by the middle of the decade the region had entered a boom unprecedented on the nation's frontiers; equally, Conceição do Araguaia was drawn closer to the national economy by the construction of the Belém–Brasília road, which was completed in 1960, and the improved extension from Guaraí to Conceição directly inspired the economic life of the region by bringing not only Belém but also Anápolis and Goiânia within reach.

For all their impact these roads did not create the frontiers under study, and were probably conceived in ignorance of them. The road to the west of Paraná answered military and geopolitical imperatives regarding the nation's security and relations with Paraguay, and it was not the mere fact of its completion in 1967 which brought the boom to the west, but rather that the road's arrival coincided with the massive economic 'investment' represented by the pioneering activity of the peasants. It was their creation of value which fuelled the boom and not the road. Similarly in Mato Grosso the road did not precede the frontier but made it possible to send frontier products to the markets of São Paulo, and in Conceição the proximity of the road precipitated but did not determine the onward movement of the pioneers (while the boom was equally the result of a wide range of Federal incentives). In short, the road catalysed the transition to commercial production.

More recently roads in Brazil have been built in response to frontier expansion, or, more radically, in a direct attempt to create the frontier by creating access to it. An example of the response is the asphalt road from São Paulo to Campo Grande in the south of Mato Grosso, which recognised the agricultural importance of the region (only 695 km of the 54,370 km of road in the state were paved at the time). The example of the attempted creation of the frontier is the proposed integration of the Amazon region through the construction of the 5,400 km Transamazonian highway from Piauí to Peru, with a second axis from Cuiabá to Santarém. The idea, as President Medici proclaimed after his visit to the drought-stricken North-East in 1970, was 'to give a people with no land, a land with no people' (*Estado de São Paulo* 1975 November). In that year perhaps one half million peasants were in danger of starvation; in the future their labour would combine with the 'riches' of Amazônia to open the frontier (as north-eastern labour had already done in Bahia, São Paulo and parts of Paraná and Maranhão). The poor of the semi-arid zones would be siphoned from the North-East and flow into the 'empty spaces' in Amazônia.

This 'impact-project' was conceived as part of the Programme of Social Integration (PIS) (Cardoso and Muller 1977), and as such was one of a range of projects (FUNRURAL, PRORURAL, PROTERRA) presented to persuade the poor peasants that they were not forgotten and to provide them, in principle, with land or succour. But it is now generally agreed that in these respects the road has failed: first built to support the penetration of an official colonisation programme, it has now been converted to service the operations of large economic enterprise; and despite the initial Federal commitment and a wide range of administrative assistance the colonisation programme collapsed. This sad story (which is seen in greater detail later in the analysis) provides convincing circumstantial evidence that the roads do not determine the expansion of the frontier economy, but rather catalyse the transition.

Agricultural production for the market

The third stage of frontier expansion always sees a rapid increase in agricultural production. In the south of Mato Grosso, which reached this stage with the arrival of the road to Dourados in the fifties, the decade brought a big rise in the output of certain crops (rice by 254%, beans by 163%, coffee by 461%); with the advance of the frontier into Pará, the cultivated area of the state increased two and a half times between 1965 and 1972 (Magno de Carvalho 1976), giving Pará 70% of the total cultivated area of the entire Amazon region; production figures for the west of Paraná (CIBRAZEM 1967) demonstrate a similar trajectory.

Table 2.5 *Production figures (in 000 metric tonnes) for five crops in the west of Paraná 1956–65*

	1956	1959	1963	1965
Rice	3	7	16	46
Beans	8	22	74	123
Maize	56	124	336	482
Wheat	30	24	11	28
Soya	1	3	13	26

But the question remains of how far this increase marks a definitive change to *commercial* production. Coffee in Mato Grosso is evidently destined for the market. But 88% of the production in Pará was still in rice, maize, beans and manioc, and only 12% in unmistakeably commercial crops (although commercial extraction continued important, especially in Marabá and São João do Araguaia); and in Paraná, it was again 'subsistence' crops which shaped the production profile. In general terms, it will be argued, in the following chapter, that such 'subsistence' crops are the principal commercial crops of the frontiers, and that the traditional function of the frontier has been to feed the cities; but it is of immediate interest to investigate what proportion as a whole reaches a market – and this will be seen to depend directly on the dominant production relations on the frontier.

Here it is necessary to distinguish between specific areas of production within the frontier region as a whole, and it is immediately instructive to divide the west of Paraná once again into south-west (below the Iguaçu river) and west (above it), and to study the evolution of the increase in production for the same crops over the same years in the two separate areas.

Table 2.6 *Evolution of production (in % terms) in the southwest and west of the west of Paraná* 1956–65

		1956	1957	1958	1959	1960	1961	1962	1963	1964	1965
Rice	SW	100	170	187	193	197	177	300	470	567	689
	W	100	248	391	687	627	382	934	1274	3099	6297
Beans	SW	100	153	280	266	399	329	373	869	1356	1382
	W	100	117	266	344	386	393	560	1434	3214	2969
Maize	SW	100	187	167	187	249	296	437	521	625	625
	W	100	98	267	442	486	496	1312	1093	1715	2361
Wheat	SW	100	79	81	76	59	43	40	32	41	69
	W	100	147	221	194	175	178	201	186	223	340
Soya	SW	100		306	408	498	679	990	1511	1145	3172
	W	100		132	275	275	276	563	537	817	1104

CIBRAZEM 1967

It is evident that (with the exception of the 'new crop' soya beans) commercial production increased much faster in the west than the south-west, where the pioneer peasants continued to produce for subsistence, sending only a small surplus to market. The approximate size of this surplus is seen in what the peasants of the municipalities of Pato Branco, Francisco Beltrão and Dois Vizinhos sent to market in the agricultural year 1966–67: 16 to 27 sacks (60 kilos) of beans; 4 to 18 sacks of soya beans; 7 to 13 sacks of wheat, and 28 to 57 sacks of maize. In other words, in the case of beans, for instance, whose production is most uniform, every peasant marketed a surplus of 0.8 to 1.5 metric tonnes.

Yet, despite the small *size* of this surplus the peasants marketed a good *proportion* of their production (with the exception of maize, which is fed to pigs) (see table 2.7).

Table 2.7 *The percentage of production sold (as against that consumed) in certain municipalities of the southwest of Paraná:* 1966

	Pato Branco	Francisco Beltrão	Dois Vizinhos
Maize	20	16	29
Wheat	31	44	36
Rice	67	36	24
Soya	51	45	71
Beans	64	80	75

Realidade Rural (ACARPA) 1966, 1967, 1968.

Any apparent contradiction here is dissolved by the small size of the farms themselves (see below) and by the fact that the peasants tended to produce the same range of crops and livestock, and did not appear to concentrate their efforts on any one product in order to place a larger surplus on the market (see table 2.8).

Table 2.8 *The percentage of farms of the south-west which cultivate or raise:*

	Maize	Beans	Soya	Wheat	Pigs
Dois Vizinho	99.7	94.3	34.8	55.2	90.9
Francisco B.	98.3	100.0	20.0	85.8	98.3
Pato Branco	100.0	94.9	43.8	91.0	100.0

Local offices of ACARPA 1966, 1967, 1968

In other words, these peasants are petty commodity producers, who send to market what is produced over and above their diverse subsistence needs. There were exceptions to this rule. Most of the maize produced, for instance, was fed to pigs, which proved to be the most profitable and most 'commercial' product: in 1966 pigs were the principal source of income of 44% of the peasants of Francisco Beltrão and of 41% of those in Pato Branco. Moreover, the peasants were prepared to experiment with soya beans, a clearly commercial crop. Their petty commodity production was commercial, but only in part.

These conclusions apply to the peasants of the south-west area, which is one of almost uniform minifundio. In contrast, production in that area of the west above the Iguaçu appeared not only commercial but highly capitalised in comparison. There were, for instance, 1421 tractors in the west at that time, compared to only 276 in the south-west; and today, in place of petty commodity producers, there are medium size farms rearing pigs commercially, and large, often mechanised farms, planting wheat and soya beans. But this contrast is not general. The distribution of those 1421 tractors was highly concentrated within certain municipalities (240 in Toledo, 230 in Palotina, 212 in Marechal Rondon, 187 in Cascavel, with the remainder spread thinly among the other seventeen) (IBGE 1970 Região Sul), and in many municipalities the production process still resembled that of the south-west. In other words, the process of production had become 'differentiated from within', leading to a *selective* penetration of the frontier by capitalist social relations.

To illustrate this it is necessary to proceed beyond the contrast between south-west and west to distinguish different production structures within the latter area. A study by Alberto Elfes (1970) provides sufficient data for this purpose. He examined two areas, the first comprising the municipalities of Toledo, Palotina, Marechal Rondon, and Santa Helena, and the second Guaíra, Terra Roxa, Assis Chateaubriand, Formosa do Oeste and Nova Aurora, and found that by 1969 they did not differ very radically in the value of the agricultural product per person employed: in area 1 it was NCr$ 598.00 (NCr$ of 1969) and in area 2 NCr$ 488.00. But the total value produced in area 2 per person employed was but 59% (NCr$ 629.00) of that produced in area 1 (NCr$ 1,065.50). This is explained in the first instance by the selection and combination of economic activities within the two regions. Pig rearing dominated agriculture in area 1, in that the traditional crops (rice, beans, cotton) had been supplanted by feed crops which occupied 82% of the cultivated land in this area, which was yet far from self-sufficient in feed which it 'imported' from area 2. Area 2 had thus become dependent on area 1 which absorbed 50% of the volume and 21% of the value of its total product (including 54% of the maize produced in area 2). So by specialising in pig rearing, which contributed 68% of the total value produced, area 1 had developed a relatively capitalised agriculture – partly at the expense of area 2. Moreover, as will be seen, the average farm size in area 1 was considerably greater than that of the minifundio predominant in area 2.

While the 'success' of area 1 was partly determined politically (much of the area fell within that of the MARIPA colonisation examined in Chapter 6) the economic differentiation of the two areas is characteristic of the selective spread of capitalism on the frontier. Further evidence to support this observation is found in this evolution of the Federal colony at Dourados in the south of Mato Grosso, where provision was originally made (1943) for 7000 farms of 31 ha minimum and 625 ha average (Diegues 1959). Of the thirty Federal colonies founded in different periods Dourados was indubitably the most successful, but this did not prevent a reversion from smaller properties to larger ranches (or *fazendinhas* as they have it in Dourados). Despite the success at different times of coffee, beans, rice and sugar (not to mention the more recent peanuts and soya beans) many of the original properties are now fused six or more together for fattening cattle. While political and administrative problems have had a part to play, economic forces, and principally the rising price of meat, which made it more profitable to sow pasture rather than rear extensively on natural pasture, worked the major changes. Once cattlemen began to put capital into clearing the forest, the petty commodity producers could not hold out against them.

Hence, in the third stage of frontier expansion, nearly all agricultural production is commercial to some degree, and some of it is capitalist. Certain instances of capitalist expansion are striking. Just as cattlemen moved to more intensive methods on higher grade soils in Dourados, so capitalist cropping moved onto the natural pastures of the south of Mato Grosso: entrepreneurs from Rio Grande do Sul, who found in 1969 that land around Ponta Porã was selling ten and twenty times cheaper (NCr$ 200 the alqueire) than in their own state, corrected the soil and mechanised production of high profitability crops like wheat and soya beans – and by 1972 were advancing on Campo Grande. Back in Paraná the same three years saw the cultivated area of Cascavél, which straddles the road to Curitiba, explode from 3000 ha to 80,000 ha, transforming it into the most mechanised municipality in the state (INCRA 1974). But however spectacular such capitalist expansion, it is still selective: capitalist relations of production are dominant but far from being universal.

The rise in land prices on the frontier

Rapid population growth, easier access to the national economy, and increasing commercial production of agricultural goods are related developments of the third stage of frontier expansion. This stage is defined by the progressive integration of the frontier into the national economy at the level of the market. In this stage value is principally created by the application of labour to land, and as soon as this value can be realised in a market for goods (in the national economy) it begins to be reflected in an emerging market for land on the frontier. In contrast to the previous two stages, title or claim to land now becomes directly convertible into capital, and land prices begin to rise.

The emergence of a market for land, which defines its price, is an economic process. Land that was abundant, inaccessible and 'cheap', is transformed into an easily reached and highly priced resource. But the regulation and institutionalisation of the market is a political process. The two processes are bound together historically, but only when the market conforms to the legal norms of the State is it correct to talk of a regime of private property. The halting and conflicted progress towards such a regime is material for later chapters (4 and 5), where it is argued that the highly imperfect implantation of the regime on the frontier is precisely what makes it politically specific, and distinct from national society. Here it is sufficient to note that, with progressive approximations to this regime, land prices on the frontier rise yet faster.

Although this rapid rise in land prices is recognised by every observer of the frontier – not to mention the legion of land lawyers and real estate agents doing business there – it is difficult to quantify these prices. Any price for land will reflect potential income from that land, and perhaps the absolute availability of land and access to it. Price of land on the frontier further reflects the uneven emergence of private property: in some areas title to land may be disputed, peasants' claims may conflict with titles, and property rights are fluid; in others the structure of land tenure may be legally secure. Moreover, even to establish income from land on the frontier (by comparing costs and value of production for certain key crops, for example) there must needs be some unit of aggregation, like the municipality; but as municipalities on the frontier continually fragment into new administrative units, and as production figures are anyway not sufficiently accurate, this order of calculation is in general not possible. But what is important to the argument are not the prices themselves but their rate of increase. Prices may vary from frontier to frontier over different periods: land sold at least ten times dearer in Paraná than in Mato Grosso in 1972, where it was expensive in comparison to Pará. However, the rates of increase are everywhere comparable; and circumstantial data for the west of Paraná can be corroborated by detailed information from certain municipalities to trace the trajectory of rising land prices on the frontier.

In the late 1950s the land in the west had not acquired a price on the evidence of the Paraná Foundation for Colonisation and Immigration (FPCI) which had to pay a commission to brokers to hawk the land in the streets of the state capital. The flat rate payable to a surveyor for mapping out an estate was 400 alqueires (alqueire paulista = 2.4 hectares). But from this time the picture began to change and from NCr$ 10.00 per alqueire in 1960 (real prices of 1972) typical agricultural land prices rose to about NCr$ 3,000.00 the alqueire in 1972 – a three-hundred-fold increase. The actual price, as suggested above, varied according to ease of access (particularly by asphalt), altitude, topography and legal status. In some places the alqueire might sell for as little as NCr$ 1,500.00 but in others the price might go as high as NCr$ 5,000.00. These selling prices can be checked for 'feasibility' by comparing them with income returned from land, where it can be ascertained: the study by Alberto Elfes (Elfes 1970) again provides sufficient data to calculate annual income from cereal growing, and total annual income, per hectare of land within the area, and per hectare of land under cultivation, for each of the two areas he considered. For the first the income per hectare of land within the area from cereal growing is NCr$ 197.00, and total income NCr$ 345.00; total income per hectare cultivated is

NCr$ 513.00. For the second the income from cereal growing per hectare of land within the area is NCr$ 283.00, and total income NCr$ 367.00; total income per hectare cultivated is NCr$ 465.00 (all incomes being adjusted to represent real prices of 1972). When we remember that there are 2.42 hectares in one alqueire, these incomes more than justify the prices being paid per alqueire.

The concentration of land-holding on the frontier

The rise in land prices is a central element in the subsequent analysis because it is this rise which heralds the struggle for land on the frontier. The rise both reflects the profit now possible from the economic exploitation of land, following the intense occupation and closer integration of the region in the third stage, and in itself precipitates speculatory purchases in search of profit in the market for land. The rise calls landowners, entrepreneurs and large economic enterprise to the frontier where they compete for land in pursuit of these profits. This competition leads to legal wrangles where their individual economic interests do not coincide, and to class confrontation with the peasants where the land-grabbing wrests their livelihood from them. This confrontation intensifies the struggle for land on the frontier which is not finished until the stamp of private property completely seals the region. The outcome of the struggle is a regime of private property which defines a particular pattern of land tenure.

As suggested above, this pattern of land tenure and land use which emerges on the frontier is the historical result of a process which is inextricably political and economic. On the one hand, political intervention through different forms of mediation is necessary to separate the peasants from the land which provides their sustenance, and to promote a specific pattern of property relations; on the other, such intervention occurs within an economic process of accumulation which defines the classes and class fractions in struggle. However, as the immediate intention is to note the final form of the effective occupation of the land in the third stage of the economic process of frontier expansion, the argument continues to abstract from the question of political mediation, and focuses on the economic forces which have shaped and are shaping the pattern. Thus, the distinction to be made between 'differentiation from within' and 'appropriation from without' does not necessarily repose on changes in the incidence and form of political mediation (still less on changes in the cycle of accumulation) but merely serves to emphasise certain relations between frontier and the national markets for goods and capital.

In the examination of petty commodity and capitalist production relations in the west of Paraná, it was already apparent that land in the south-west corner of the region (below the Iguaçu river) is divided into relatively small peasant plots: in fact in 1965 74.7% of the holdings were smaller than 25 ha and 92.7% smaller than 50 ha (CODEPAR/SERETE 1965). By Brazilian standards this makes the region one of minifundio, but even here the holdings smaller than 50 ha account for only 60% of the total farm area. Moreover, above the Iguaçu, the pattern of tenure is far less uniform, as is clear from a further comparison of the two areas of the Elfes study (in percentage terms which throw the different patterns of tenure into relief).

Table 2.9 *Differing patterns of tenure in two areas of the west of Paraná*

	Area 1		Area 2	
Size of farm	% of no.	% of area	% of no.	% of area
up to 10 ha	24.2	6.3	40.0	28.0
10–20 ha	25.6	11.3	39.0	29.8
20–50 ha	42.4	45.8	19.8	36.0
50–100 ha	5.7	18.8	1.0	4.0
100 ha+	2.1	17.8	.2	2.2
	100	100	100	100

Alberto Elfes 1970

While the incidence of minifundio is yet higher in the second area than in the south-west (farms smaller than 20 ha accounting for 79% of the properties and 58% of the area), in the first area it is the medium size farm which predominates (48% of the farms and 65% of the area), and this difference is reflected in the relative population densities of the two areas – 38.5 inhabitants/km² in the first area and 60.2 in the second (the number of people employed in agriculture in the two areas is roughly the same). In short, within the relatively equal pattern of peasant land-holding, the figures indicate an incipient differentiation between smaller peasant plots, or minifundio, and larger, more capitalised properties.

Similar tendencies can be observed in the case of Mato Grosso. At first glance the advance of the frontier seems to have had 'favourable' effects on the pattern of land tenure in the state. Over the decade 1950–60, for instance, the number of rural holdings increased by 201% while the occupied area increased by a mere 11.7% (to cover

26.3% of the total area of the state), so reducing the average property size from 1812 ha to 672 ha (IBGE census data). In the Campo Grande region, which accounted for 51.3% of the cultivated land in the state (Dourados alone accounted for 16.3%) the average size fell from 955 ha to 311 ha. But an assessment of the overall change in distribution of land holdings gives a different picture.

Table 2.10 *Distribution of number and area of land-holdings in Mato Grosso by size of property 1950–60*

	Number		Total area (ha)	
Size	1950	1960	1950	1960
up to 10 ha	8.6	23.3	0.02	0.20
10–100 ha	29.92	43.38	0.61	1.95
100–1000 ha	34.59	17.45	7.54	9.52
1000–10,000 ha	23.12	9.51	38.38	42.62
10,000 ha+	3.75	1.43	53.45	45.71
	100.00	100.00	100.00	100.00

The distribution curve sharpens with the concentration of medium (100–1000 ha) and large (1000 ha+) properties at one end of the scale, while small-holdings, at the other, tend to fragment (the increase in the number of these, 460%, being greater than the increase in their total land area, 278%). In short, the process of differentiation in Paraná is evident again as a pronounced tendency to accentuate the dominant division of the countryside into minifundio and latifundio.

In economic terms this tendency can be traced to the integration of the frontier into the national market economy, and to changes in demand in that economy. A full analysis of the expansion of the frontier in terms of the market must await the next chapter; at this moment it is only possible to suggest, synoptically, the way in which wholesalers and industrialists begin to compete with local buyers for cereals and beef, and so contribute to the opening up of the market and the breakdown of the oligopsonistic structures of the regional centres. This raises prices paid to the producer. But the large wholesalers from the cities do not stop here but move on to buy direct from the producer in order to ensure a steady supply (Forman and Riegelhaupt 1970a and b) (and the producers prefer to sell direct to the wholesaler at lower cash prices than to the middleman on credit). The atomised supply typical of a regime of minifundio cannot satisfy this direct demand pull, while, on the demand side, the wholesalers tend to favour large producers in order to lower costs (and by buying in bulk they can pay higher prices). Therefore, properties begin to fuse (which paradoxically will occur most rapidly in areas of larger properties where the wholesalers are already buying) and the higher prices paid to bigger producers make capital investment, or investment in more land, attractive. In short, the new 'funnel-like' flow from producer to industry (or, more recently, along 'export corridors') through the large wholesalers excludes both smaller intermediaries and smaller producers: capitalisation of the commercial sector leads to capitalisation of production, which raises land prices high enough to exclude the small producer from the market and so concentrate land-holding. In this way the 'select areas' mentioned above are created, and the process of 'differentiation from within' pushed forward.

But the differentiation in the distribution of land in response to demands for goods in the national market is not alone in determining the duality of land-holding in the third stage. The characteristics of the national market for capital are also important. Traditionally, the frontier has attracted speculatory capital requiring rapid and high rewards; this tradition has been reinforced by the recent range of fiscal incentives which – in combination with incentives to exports, and especially beef and grain exports – have succeeded in transferring a mass of money capital to the frontier, where, by definition, the organic composition of capital is low. The 'frontier projects' are designed to maintain high average rates of profit in the economy as a whole, and create an expanding market for the capital goods sector, whereby fiscal incentives should result in real investment, making financial accumulation compatible with real accumulation (Oliveira 1972a) (and any increase in exports will earn the foreign exchange to pay for oil and technology imports and service the massive foreign debt). The impact of the capital market is such that it can change the very structure of production on the frontier and in so doing concentrate land-holding by a process of 'appropriation from without'.

Evidence for this is found in Pará, where the whole economic base of the state was transformed between 1960 and 1972. In 1960 there were approximately 66,000 properties engaged in agriculture and cattle-raising; by 1972 this figure had dropped to 36,000, but in comparison with the 2,200,000 ha occupied by all the properties in 1960, the 14,000 uniquely engaged in cattle-raising in 1972 occupied over 7,000,000 ha (Pinto 1976b). In this year cattle-raising had taken over some 40% of the total occupied area of the state. Moreover, the number of holdings practising extractive activity fell from 12,000 to 571 over the same period, and there are suggestions that the profits from the few remaining extractive enterprises, especially cashew nuts, were going into cattle.

This remarkable turn-about in the economic activity of the state explains the almost equally remarkable concentration of landholding. Between 1960 and 1972 the number of properties decreased from 83,000 to slightly fewer than 40,000, while the area occupied increased four times over to 20,909,690 ha (INCRA 1974). Of these properties in 1972, those smaller than 500 ha accounted for 88.5% of the total but occupied only 12% of the area; while, logically, those bigger than 500 ha comprised 11.5% of the total and occupied 88% of the area. Moreover, everything indicates that the number of small properties actually decreased over the period. In 1960 71,432 of the 83,000 properties were smaller than 100 ha, while in 1972 INCRA categorised as minifundio (a minifundio in the Amazon may be bigger than 100 ha by INCRA criteria) only 28,004 holdings. In 1972 these minifundios occupied only 1,200,000 ha of the total occupied area, the rest being taken by latifundia and what INCRA calls 'rural enterprise'. A further striking index of the extensive nature of this occupation was the 75.4% of the occupied area which is unexploited economically, against the figure of 42% for Brazil as a whole (even if half of this 75.4% can in principle be counted as 'legal' forest reserves). These dramatic changes in the pattern of land tenure were achieved through 'appropriation from without' by entrepreneurs and economic enterprise.

The emergence of a 'free' labour market

The question now arises of the economic destiny of the peasants formerly occupying the small-holdings which, in Pará at least, were fast disappearing. The answer is suggested by a closer look at this economic transformation as it occurred in Conceição do Araguaia in the south of Pará. There in 1972 the 646 minifundios (by INCRA criteria) composed 40.2% of the properties and 2.54% of the occupied area, while the latifundios accounted for the other 58.8% of the properties and occupied 97.46% of the area. But this sudden expansion of cattle-raising estates took place in a region which had supported a largely subsistence peasantry since the rubber decline and which had felt the migratory impact of the pioneer frontier. The number of minifundios simply seems too small to account for the population existing in the municipality. Certain small townships had developed, it is true, but the majority of the population worked in the countryside. The only place to look for this population is in the cattle estates themselves. In 1960 these employed a mere 144 permanent and 209 temporary wage-labourers, and the picture was little changed in 1970 with figures of 183 and 218 respectively. But by 1972 the estates employed 857 permanent and 7011 temporary wage-labourers (Ianni 1977) (and perhaps others who were not 'declared'), who formed the new 'free' labour force of the municipality. In all of Pará in that year some 17,000 wage-labourers were employed overall in the countryside, with as many as 40,000 temporary workers taken on at any one time (INCRA 1974).

These figures reveal the strikingly high proportion of 'temporary' to permanent wage-labourers on the cattle estates. Such labour is largely engaged in the work of clearing the land of forest (*a derrubada*) and putting down pasture. Traditionally recruited from the North-East this labour has been increasingly drawn from the ranks of the dispossessed peasantry on the frontier itself. In both cases it is the company 'contractors' (*empreiteiros* or *gatos*) who recruit the labourers, and conduct them to the work place, where their work is perennially 'temporary'; once the clearing is over their labour is no longer required. Traditionally, again, these labourers might have been allowed to plant rice or subsistence crops for anything from one to three years (*troca pela forma*) to defray the costs of clearing on the estate, but even this practice is falling into disuse. Contemporarily, there are few options but to await the next 'contract', push on to the next frontier, or revert to subsistence – which, with increasing monopoly of land, is relatively rare. It is estimated that in the whole of the Amazon region there are as many as 200,000 of these 'temporary' workers, or *volantes*, living in conditions of desperation (*Estado de São Paulo* 1975).

This massive rural 'reserve army' of labour is in part the creation of State fiscal and credit incentives to economic enterprise in the Amazon region (SUDAM 1976a) which have subsidised the investment of capital, but not the payment of wages. As capital is cheap, these cattle estates are highly capitalised enterprises, with a very low ratio of employment to capital invested: it is calculated that on average one wage is paid for every 256 cattle. No more, it may be mentioned, do the incentives subsidise production, and the product–capital ratio is also low. At the same time these enterprises are highly extensive, and require large areas of relatively fertile land for their reproduction; the average size of a cattle ranch in Amazônia is of the order of 18,750 ha (Pinto 1976b). In short, the cattle estates monopolise more and more land, especially after the widening of the incentives in 1969, and employ less and less labour. Almost paradoxically this would not be a viable pattern of expansion for the large estate were there not so much labour 'available' for temporary work; but it is the estate itself which creates this 'reserve army' by the monopoly of the land and the dispossession of the pioneer peasants. These are thrown into a 'free' labour market only to find that they are only 'free' of any possible means of support.

Capitalist production on the frontier

During the first two stages of frontier expansion accumulation does not take place primarily through the appropriation of surplus-value, but in the third stage properly capitalist social relations begin to dominate the frontier economy. This does not deny that the whole process of expansion is simultaneously one of integration into a capitalist market: values created outside capitalist relations enter a sphere of circulation subject to industrial capital – and this is the argument of the following chapter. But only in the third stage does the capitalist enterprise emerge at the centre of the economic contradictions and political struggles on the frontier. In the case of Pará, the cattle which grow fat on the land lost to the peasants create the conditions of direct antagonism between these peasants and the cattle-raising bourgeoisie. The ensuing struggle is the social expression of the historical contradiction between retrenched petty commodity production and an expanding capitalism.

The transition to dominant capitalist relations on the frontier is determined by the national capitalist economy. Contemporary developments in this economy have tended to speed up this process of capitalist penetration. In the first place the preponderant presence of international monopoly capital has greatly increased the operational capability of capital on the frontiers of the economy; in the second place the State has begun to intervene more directly in this process. Not surprisingly, the State has intervened decisively on the side of capital. As will become evident, it had always been present on the frontier to mediate legal and political struggles and oversee the definition of private property; but its present incentives schemes, for instance, actively work to concentrate property under the control of large capital and so drive the peasants off the land. But such State intervention is not a 'mission' to make capitalism 'victorious'; the State responds to objective constraints which are the contradictions and crises which threaten capitalist profit. In other words, there is no more reason contemporarily than historically to suppose that the transition to capitalist production on the frontier is ever complete.

Looking at the question empirically, it is not necessary to rehearse the many respectable arguments (Paiva 1966; 1968; 1971; Nicholls and Paiva 1965a and b) against the economic possibility of a generalised use of modern technology in Brazilian agriculture; not only in economic but also in social and political terms it is difficult to conceive of a primary sector in Brazil which employs 10% rather than the present 50% of the economically active population. On the contrary, the transition to capitalism has occurred and will continue to occur within 'select areas' like those observed in the west of Paraná, the south of Mato Grosso and the south of Pará. In these areas capital is invested and wages paid – at least occasionally. As the market for land becomes institutionalised the capitalist owner requires a return on investment at least equivalent to that on capital invested in industry (although the incentives play a part here). Outside of the 'select areas' the land and labour not drawn into capitalist production either combine in extensive and subsidiary forms of production or cluster in reserve areas which may even cease to produce for a market, except on an occasional basis, but may well continue to supply 'temporary' or 'seasonal' labour (*volantes; bóias-fría*) to the other sectors.

Within a more theoretical focus, a teleological perspective which proposed an increasing homogeneity of capitalist production and market relations might project a highly capitalised sector, where increasing demand brings further technological advances and inputs, with more and more labour and land 'dumped' in residual areas. In such a perspective only capitalist production contributes to accumulation. The assumption here, on the contrary, is that different production relations contribute to the creation and appropriation of value, and through the continuing transfer of this value participate in the overall process of accumulation. In other words, accumulation in this capitalist economy is achieved not only through the appropriation of surplus-value but through the transfer of values from sub-capitalist modes of production which are *articulated* to the dominant capitalism. A fuller analysis of such accumulation on the frontier and its significance for the reproduction of the economic system as a whole must await the final part of the argument (Chapters 8 and 9). The assertions are advanced here only to indicate that the increasing penetration of capitalism on the frontier does not complete a transition to capitalism. More labour moves to the frontier, more land is taken into production and more capital is invested; but what emerges from the increasingly intricate combinations of these factors is not universal capitalism but a new form of the continuing articulation, achieved at a different level of development of the productive forces.

Throughout this account of the transition to capitalism on the frontier, which is simultaneously a process of integration into the national economy, the argument has assumed the determining presence of this economy. In particular, it has assumed certain characteristics of the national market for labour, which propel migration to the frontier, and of the national market for goods, which promote frontier production. A further assumption has surfaced regarding the transfer of value from the frontier to the national economy. In the following chapter this relationship between frontier and national economy is made explicit with the intention of discovering the motor of migration, and describing how market mechanisms contribute to this transfer.

3

Frontier expansion and the national economy

Frontier expansion was interpreted in the last chapter as a process of integration into the national economy. This integration was seen to proceed through interrelated changes in both production and market relations: the integration which occurs through the progressive change to commercial production in response to the national market is simultaneously a transition from non-capitalist to capitalist production. This interpretation illuminates the process itself, but does not yet pose the question of the relationship between frontier and national economy; in effect, it suggests *what* occurs in frontier expansion but not *why* it occurs. Nevertheless, in proceeding to answer this latter question, it is apparent that the prior investigation of the process has revealed the principal dimensions of the determining structure – at least in the economic realm. Thus the question of the relationship between frontier and national economy is made two-fold: on the one hand, what are the production and property relations which determine frontier expansion and accumulation; on the other, what are the market relations which determine the transfer of value from frontier to national economy? In no sense are either or both sets of relations, which in the reality are not easily separated, to be taken as uniquely 'determining': their separate investigation here is simply intended as one step in the presentation of the political economy of the frontier.

This step aims to establish the structural conditions of frontier expansion, and so rejects any idea of 'natural or ideal determination' of the process. To look first at the level of production, it is therefore production and property relations which determine the expansion, and not population growth in itself, which has no meaning outside of the structure of production of a specific social formation. What is important is to establish this specificity, which is at the same time the specificity of the pioneer frontier, and investigate what happened historically both in city and countryside to create the conditions for expansion. Thus the analysis attempts to outline the conditions which emerge from the process of industrialisation in this century, and the changes in the markets for labour and staple goods, and assess their effects on frontier expansion. Similarly, at the level of the market, it is not any and all exchange relations which achieve the transfer of value from the frontier, but historically specific market mechanisms, which vary with the degree of integration of the frontier into the market, and with the structure of the market itself. Thus the specific assumption here is that the frontier is in the third stage of its expansion, and, at the same time – as the market is not ideally 'free' or 'undetermined' – an attempt is made to assess the impact on its structure of the political interventions of a particular period – which is central to the overall period of the pioneer frontier. Later in the presentation of the political economy these restrictive assumptions are removed, and the analysis of the transfer extended through all stages of expansion, and into the contemporary context.

Pioneering and industrial growth

Writing in 1935, J. F. Normano claimed to have coined the expression 'the moving frontier' and proceeded to describe all of Brazilian development in terms of the economic occupation of the half continent which is Brazil. 'The expansion of Brazil had and has a pronounced internal character. It is a process of putting an economic substance in the political area, of bringing the economic nearer the political frontier. This movement formed the nation' (Normano 1935 p. 2). His description emphasised the country's ability to expand production extensively in rapid response to new demands in the world market. This ability depended on the massive availability of land and gave Brazil shorter or longer periods of supremacy in different 'new products' (sugar, cotton, cacao, gold, rubber, coffee). These periods correspond to the 'booms' which have given to Brazil its peculiarly 'cyclical' development.

In general terms, Normano's definition of the frontier will be seen to be similar to my own, but his interpretation of the frontier is finally much broader. He is concerned to demonstrate the response to new demands in the world market – through what he calls 'a feverish enlargement of production' through an extensive expansion – and therefore it appears that any new economic activity in any part of the national territory, occupied or not, can be classified as a frontier experience. For my part, I shall interpret the frontier for most of its history as a response to domestic demand (and the 'demands' of domestic accumulation) generated within the national economy. However, what seems at first sight to be a clear conceptual distinction may simply depend on different historical perspectives, and the changes in the Brazilian economy which began to take place precisely at the time Normano was writing.

There are two principal changes. In the first place, Brazil has experienced a rapid process of industrialisation since the 1930s. This industrialisation transformed the previous pattern of cyclical and regional growth and, in effect, there have been no new agricultural 'booms' since 1935 (though coffee has suffered broad oscillations on the world market and cattle raising for export has expanded greatly in recent years). It contributed to the high rate of urbanisation over the period and finally to the clear dominance of the urban economy over the countryside. As part of the process, and important to the frontiers, it greatly expanded the internal market and radically altered the pattern of demand (Baer 1964; 1965; Baer and Maneschi 1969). Despite all these changes, Brazil continues to depend, like most third world countries, on the revenue from the export of primary products, and principally coffee, but the export economy is no longer central to the country's capacity to accumulate (Martins 1969a). In the second place, the Brazilian economy began at the same time to exhibit signs of a labour 'surplus'. Normano had not only insisted in his analysis upon the 'immense reserve of land for expansion', but implicitly recognised the scarcity of labour on the 'moving frontier', pointing for instance to the movement of labour from the North-East to the gold fields of Minas Gerais as the principal cause of the decline of the sugar economy in the eighteenth century. This scarcity continued into the twentieth century and the continuing eagerness of the country and especially the state of São Paulo to import Italian labour bears witness to this. But the situation changed dramatically, and of course the very expansion of industrial production must also have contributed to this second change by its partial destruction of petty commodity production (especially artisanal production) in city and countryside (Castro 1969; Singer 1968). In my view these two crucial changes in the domestic markets for labour and staple foods were sufficient to change the whole significance of the 'moving frontier' in Brazilian development.

The industrialisation contributed to the concentration of population and to the increase in demand for staples. The labour 'surplus' was available to move onto the frontier and meet this demand. In fact, that agricultural production has grown as fast as population over the last 40 years and has responded rapidly to the changing demand conditions created by the industrialisation is largely due to the frontiers. For this growth has continued to be extensive through taking more land into production and until very recently there has been little or no increase in productivity and no change in the social relations of production on the frontier (Paiva 1966; 1968). This pattern of growth could continue, firstly because land was abundant and secondly because the labour surplus soon became an 'excessive surplus' (F. H. Cardoso 1975) and there was consequently no need for technical innovation in agriculture in order to free labour for industrial enterprise (Nicholls and Paiva 1965b).

These broad assumptions about the relationship between frontier and national economy compose the premises of the present argument, and it is necessary to establish at the outset that there are reasonable grounds for accepting them as true. In other words, this interpretation of the relationship must be seen to fit the available facts. But the statistical data on the diverse aspects of the economic growth of Brazil over the period vary widely in quality and quantity, and selective reference to the data has provided support for different views. It is therefore important to refer the assumptions to a consistent body of data, and, for the most part, the references throughout the chapter are made to Kahil's book (1973) on inflation and economic development in Brazil. Kahil's work was chosen in the recognition that his data series are probably the most skilfully composed and statistically sound; and as his meticulous account is 'monetarist' in its convictions, there can be no danger of confirming the assumptions by a 'selective bias' of the data.

The first assumption is simply that agricultural production in Brazil has grown as fast as population over the last forty years. There are no satisfactory series for the whole period, but a comparison for the years 1950–64 between the rates of growth in per capita agricultural production and in per capita availability of foodstuffs (Kahil p. 50) clearly demonstrates that increasing provision of food to the population exactly accompanied agricultural growth. Furthermore, it appears that this may partially have been achieved through the changing overall composition of agricultural production. A second series comparing the rates of increase in total crop production and in crop production for internal consumption over the years 1947–63 (Kahil, Appendix 4) shows the total to have increased by approximately 85% over the period, while production for internal consumption more than doubled.

The corollary to the first assumption is the claim that this successful growth of agriculture in response to increasing internal demand is due to the frontiers. The difficulty in testing this claim is that it appears impossible to disaggregate the data on agricultural production in such a way as to show what proportion of the total production of the sector has been produced on the frontiers. However, there are figures for agricultural crops over the period 1937 to 1961 which compare their real prices to the producer, their output, and, crucially, the area sown to these crops (Kahil p. 169), which provide strong indirect support for the claim. Kahil divides the twenty-four years in question into six four-year sub-periods for the purposes of presentation. As expected, and despite a net decrease in real prices, crop output increased enormously over the latter part of the period: after remaining more or less stagnant over the first eight years, output increased by over 10% in the two following sub-periods, suddenly registering a massive 40% rise between 1953/54 and 1957/58, and falling again to 25% over the last four years. Significantly this increase in production was almost matched by the rate of expansion of the area planted to crops, which ran well ahead of production over the first sixteen years (registering −4.9%, 19.4%, 12.7% and 16.7% for the four sub-periods respectively) only to fall behind over the last two sub-periods (16% and 17% respectively). This appears to indicate, at the very least, that a large proportion of the increase in production was achieved precisely by taking more land into production.

Finally this claim can be confirmed by implication if it can be proved, as asserted above, that little was done to raise productivity in agriculture. In this connection, Kahil provides information on the changes in average yields per hectare of Brazil's main agricultural crops over the years 1947–61 (Kahil p. 126; his fifteen crops occupied about 85% of the total cultivated area and their value represented 90% of the total). Remarkably, there was no change in the average yields of beans and manioc, while that of rice rose by 1.0% and that of wheat fell by 20%. Yields of both maize and sugar increased by 9.0%, it is true, but overall productivity in the traditional staples remained more or less constant. In direct contrast to this part of the picture, yields per hectare of the principal cash crop, coffee, rose by no less than 87%. The exceptional performance of coffee, however, which goes predominantly to export, only serves to reinforce the main line of the argument, that the increase in the production of staples in response to rising urban industrial demand has been accomplished through the advance of the frontiers. In a word, the frontiers have done most to feed the cities.

The rice, maize, beans, manioc and pork eaten in São Paulo and Rio de Janeiro have been produced on the frontiers in Rio Grande do Sul, the west of Santa Catarina and west of Paraná, the south of Mato Grosso and south of Goiás, the 'Minas Triangle', and the west of São Paulo. Advancing within a radius of anything from 500 to 2000 kilometers from these cities the frontiers have accompanied the rhythm of Brazil's industrialisation, and its capitalist development (Velho 1972). This advance accelerated after the Second World War in answer to the intense pace of this industrialisation, such that contemporarily – as was evident in the last chapter – some frontiers have even penetrated the Amazon region. As they have advanced these frontiers have been progressively integrated as forecast by Monbeig (1952) by those 'relations sans cesse plus étroits . . . entre les industries des grandes villes et la frange pionnère'.

Thus this study of the frontier in Brazil begins where Normano stopped. Over the period of history covered by his analysis, producers for the internal market constituted an important but 'unnoticed' sector in agriculture, who produced in the shadow of the dominant export economy (Cardoso 1975). They now step out of the shadow and, by pushing back the frontier, develop the agricultural economy in breadth – in response to the in-depth development which takes place through capitalised work relations in the industrial and financial centres of the national economy. Moreover, the fact that they do this on the frontier is important. The prices of staple foods have remained low, precisely because they are produced in frontier areas of high productivity: rudimentary techniques may predominate in both old and new areas but the fertile soils of the frontier guarantee high yields, at least in the short term. So it is proper for the study to proceed immediately to investigate the phenomenon of migration to the frontier. Without this migration the modern form of frontier would never have appeared.

Migration to the frontier

Traditionally, migratory flows have responded to the cyclical pattern of growth presented by Normano. Hence in each period they tended to converge on one region. First came the occupation of the coast and the extraction of pau-brasil; next the cycle of sugar monoculture in the North-East followed by the gradual penetration of the sertão to rear cattle. Then came the descent to the mines of Minas Gerais and, towards the present century, to the coffee plantations of the high plateau of São Paulo. Evidently the overall direction of the flow is towards the south but the significant interruption of the tendency during the rubber boom in Amazônia demonstrates that this is accidental and that Normano's insistence on the pull of the world market is correct. What is traditionally true is that it is the North-East which supplies most of the migrants. In the case of rubber the population of Amazônia grew from 340,000 in 1872, the beginning of the boom, to 700,000 in 1900 and 1,400,000 in 1920 (Tupiassú 1968). By far the greatest part of the increase is due to immigration and it is estimated that at least half a million north-easterners entered the region during those years. They were not merely drawn out by the boom but driven out by the droughts which repeatedly struck the North-East in those and following years (Hirschman 1963).

In recent decades, although the North-East may finally supply a majority of the migrants overall, the migratory flows have been far more diverse and have progressively increased both in scope and intensity. According to the census data 8.5% of the national population resided outside their state of origin in 1940, 10.3% in 1950 and 18.2% in 1960 (Graham 1969): that is, in each of these decades this proportion of the population have moved home at least once in their lives (and as census data only recorded migratory movements which cross state boundaries it is apparent that the real mobility of the population must be considerably higher than that indicated in these figures). In 1970, following refinements in the census, it was discovered that over 30,000,000 Brazilians out of roughly 100,000,000 live in municipalities other than the one they were born in. The majority of these migrants have moved from the countryside to the towns, often drawn directly or indirectly by the centripetal force of São Paulo; the rest have moved in a contrary direction towards the constantly advancing frontiers of the interior.

The dimensions of this movement during the fifties can be gauged by the increase in the economically active rural population in the states with expanding frontier regions: from 300,000 to 500,000 in Goiás, from 400,000 to a million in Maranhão, and from 500,000 to 1,300,000 in Paraná (Ribeiro 1970). Figures on overall population increase are no less expressive. The population of Paraná, where the frontiers advanced both in the north and the west, grew from 1,240,000 in 1940 to seven million in 1970, and the west of the state, only sparsely populated in 1950, had 1,200,000 inhabitants by 1970 (IBGE 1970 Paraná). Similarly the population of Mato Grosso, as noted in the last chapter, rose from 520,000 in 1950 to 1,625,000 in 1970, with Campo Grande in the frontier region of the south accounting for some 40% of the total increase (Coelho 1971; IBGE 1970 Mato Grosso). Finally, in Pará, where the advance of the frontier is most recent, some 392,000 of its 2,200,000 inhabitants in 1970 had been born outside the state (IBGE 1970 Pará); over the previous decade alone some 130,000 migrants, mainly from the North-East, had moved into the south of the state (Magno de Carvalho 1976). In short, this 'counter-migration', or pioneering, intensified as rapidly as the major movement to the cities, and may equally be considered a response to changing economic conditions.

These migratory movements in the south and centre-south of the country – in Rio Grande do Sul, Paraná, Mato Grosso, Goiás – are understood to be largely 'spontaneous' insofar as the millions of migrants usually arrive at the frontier through their own resources, exiguous as these may be, and by their own 'volition'. The true nature of this 'spontaneity' is examined below in the context of the economic constraints acting on the migrants; but in general it remains true that no political initiatives by the State have been necessary to encourage the majority of the migrants to move to the frontier and settle. Where the State has acted at all it has usually been in reaction to the migration – intervening in an attempt to resolve conflicts and contradictions arising out of the movements – and this observation is supported, with minor qualifications, by the later discussion (Chapter 6) of local state and Federal State initiatives on colonisation. However, State intervention on the frontier in Amazonia appears to provide an exception to the rule.

In recent years the State has intervened 'ahead' of the migratory movement in Amazônia in order to 'open up' the region and direct the process of colonisation and programme of investment. But, as noted briefly in the last chapter, the investment rapidly took priority over the colonisation. The most dramatic State initiative was the construction of the major penetration roads into the region (Belém–Brasília, Transamazônica and Cuiabá–Santarém), which constituted the essential infrastructure of the Programme of National Integration; but the collapse of the sponsored colonisation schemes and the increasing fiscal and credit incentives to national and international enterprise led to penetration by monopoly capital not people (Cardoso and Muller 1977). While none of this denies an indirect stimulus to 'spontaneous' migration, the State initiatives must largely be understood as providing supports for capital projects quite beyond most private resources – for instance in the exploitation of iron ore in the mountains of Carajás or in the mining of bauxite in Trombetas. And, as will be argued in Chapter 7, this reflects the growing dominance of monopoly capital in the economy as a whole and the close links established between capital and State bureaucracy in the contemporary period. But these changes should not obscure the unchanging determinants of frontier expansion. These contemporary developments do not constitute a return to the former pattern of migration which pursued regional 'booms' – although there is no doubt that the State has contributed to engineer a 'boom' in Amazônia; the frontier in Amazônia, on the contrary, is an extension of the pioneer frontier, and conforms to the same cycle of accumulation and the same motor of migration.

The motor of migration

The migration may be a more or less independent variable in the analysis of the occupation of the land but it is itself determined by social and economic structures in the countryside which reflect the total economic formation of Brazil. In this connection it is inadequate to seek to understand the migratory movements in terms of the individual motivations of the migrants themselves. Just as the concept of social mobility in social science is a poor euphemism for individual mobility, so debates over why individual migrants migrate, or over which migrants from a given population migrate, obscure the social determinations of the movements. Similarly, 'theories' of migration which propose a migration 'system' where two areas interact and certain typical actions come to define the process (e.g. Pastore 1969) are plainly empiricist and, as such, cannot discover cause and effect – let alone predict future developments. All recorded motivations for migration such as the 'lack of land', the 'lack of work', and 'end of the contract', the 'end of the harvest', the 'badness of the boss', the 'hunger', the 'drought', the 'declining fertility of the soil', the 'hate of hiring their labour out' reflect the economically inefficient and socially unjust combination of land and labour in Brazil, whereby the great mass of land is monopolised in a few hands and the millions of peasants denied access to it (Chacel 1969; CIDA 1966). This monopoly causes certain predictable effects for those who have no way of producing their subsistence other than by working the land. As Juarez Brandão Lopez puts it, 'the imbalance between population and the means of subsistence is what determines the abandonment of the land' (Brandão Lopez 1964). The migrants move to the frontier in order to survive. The movement is spontaneous in that they are looking for land of their own to work. But the failure to be 'spontaneous' means starvation.

These assertions take on substance in the examination of any area of 'out-migration' in Brazil. The state of Rio Grande do Sul, for example, has supplied the migrants who over a period of decades have peopled the west of Santa Catarina, the west of Paraná, the south of Mato Grosso and who today are moving into Rondônia in the Amazon region. The first cause of this is the land distribution within the state which is seen to be highly unequal (but not more so than elsewhere in the country) (see Table 3.1). By a different survey four years later (INCRA 1971) farms of less than 25 ha were seen to be 73.9% of the total number but covered only 18% of the total area in farms. Within this overall pattern of concentration two areas of extreme minifundismo or farm fragmentation can be distinguished: Alto Uruguai,

Table 3.1 *Distribution of land in Rio Grande do Sul*

Size of property	% of total no. of properties	% of total area
up to 10 ha	35.2	3.9
10–100 ha	57.8	19.1
100–1000 ha	6.4	49.1
1000 ha+	0.6	27.9
	100	100

Departamento de Cadestro e Tributação do IBRA 1967

with 111,825 farms, and the Encosta Inferior do Nordeste, with 88,920 farms. These areas are known as the old colony and the new colony and contain about half of the 'small farms' (less than 50 ha) within the state: the number of these increased 438% between 1920 and 1967 but the area they occupied increased by only 255%; the average area of the small farms decreasing from 24 ha to 15 ha over the same period. In short, according to the criteria of INCRA nearly every farm (95%) in these two areas is minifundio – that is, it cannot provide subsistence for the family which works it. These figures clearly demonstrate an increasing pressure on the land in specific areas due to a high secular rate of demographic growth (3.8% per annum in the colonies as against 2.6% in the state as a whole) and to a lack of access to land elsewhere in the state.

The pressure on the land is aggravated by the Brazilian laws of inheritance, whereby, on the father's death, half of his land goes to his wife and the remainder is divided between his sons (Código Civil Brasileiro). The law encourages the progressive division and subdivision of the land as generations come and go, which means that the minifundio areas must support an ever-larger population at a decreasing standard of living. In 1967 there were 434,956 minifundios in the state (not including the 28,232 renters of different sorts, and the 66,414 share-croppers – IBRA 1967b). And it is estimated that some 30,000 families enter the labour market each year. It is therefore not surprising to learn that according to INCRA's calculations (1971) the underemployment in the new colony and the old colony runs as high as 45 and 47% respectively. These figures represent the percentage of the total potential labour force not employed and are related to the overall distribution of land in the state in order to demonstrate that what exists here is a 'structural labour surplus'. This huge labour surplus has little chance of finding work. On the one hand employment in agriculture, which absorbs about 50% of the economically active population of the state, expands at a mere 0.9% per year, and, on the other, migration to the cities cannot be considered a viable alternative: industry cannot take anything more than a tiny proportion of the migrants while the tertiary sector which grew from 20–36% of the active labour force from 1940 to 1950 is now 'saturated'. So the surplus may stay and hope for seasonal employment in capitalist enterprise in the countryside, or migrate to the frontier.

Finally the migrant moves when the farm can no longer provide subsistence for all members of the family (he may have inherited a small plot which will not support his children or may himself be a burden to surviving parents farming impoverished soils). Whatever the labour input, this point will inevitably come as the fertility of the soil inexorably declines. The more minute division of the land in the west of the Alto Uruguai where the soils are richest (plots of less than 10 ha cover 22% of the cultivated land and the majority of farms are smaller than 15 ha) indicates that the size of the farm is progressively reduced to what 'does to live off'. When this point is passed 'spontaneous' migration begins. Strangely little seems to be done to slow the decline in fertility. The original German and Italian immigrants to these colonies soon reverted to the local *sistema de bugre* (Roche 1959) (rudimentary farming which exhausts the soil) but even in recent years the same land is farmed year after year (without rotation or fertilisation but with ever higher labour inputs) until it can 'give' no more (Erven 1969; Sá 1968). But the lack of investment to slow the decline is not surprising in the light of the average ratio of the value of annual product to available liquid capital on farms of 20–30 ha which varies from 49–155% averaging at 91% (Prebayle 1967). In these conditions, after subsistence costs are met there is little capital left to invest: the peasant tries to achieve the best production possible in the shortest space of time with the smallest possible outlay. And, very logically, his first preference for investing this meagre capital, again according to Prebayle, is in new land on the frontier. He will usually move after the summer harvest; on arrival he will have the winter months to clear the first of the forest.

The small farmers were priced out of the land market in Rio Grande do Sul from the early 1950s, which saw a rapid capitalisation of production in crops such as wheat, soya beans and rice (increases in the area planted to these crops over the decade were 99%, 53% and 34% respectively) (Cano de Arruda 1967). Other areas within the state were monopolised by cattle raising and the coastline was already crowded. There was no choice but to move on to the virgin soils of the west of Santa Catarina and south-west of Paraná: the only good news was that land prices in these regions were as much as four times lower than in Rio Grande as late as 1964. Thus the tempo of out-migration from the colonies has steadily increased. Since 1958 anything from 100 to 400 families have left each municipality each year. These families are squeezed out of the state and forced to the frontier. And after some few years the frontier reaches the same point of saturation. In the south-west of Paraná the economically active population increased from 38,000 to 155,000 over the period 1950–67; some 20,000 new labourers entered the market each year (INDA 1969). Thus the same pressures are reproduced and the migrants pushed out into the west of Paraná and south of Mato Grosso. This is the – self-generating – motor of migration.

The appropriation and transfer of value from the frontier

It has been said that the growth of agricultural production in recent decades has been extensive and achieved by a continually renewed combination of land and labour. This is as true today in Amazônia as it has been for the expanding frontiers of the south. Almost paradoxically, the proportion of the land in Brazil which is cultivated is relatively small, given the size of the country and the number of inhabitants. It is indicative that in 1960 Brazil, with a far larger population than Argentina (70 against 21 millions), actually cultivated less land (29,000,000 ha against Argentina's 30,000,000). In that same year the average area cultivated per head of the economically active rural population was but 2 ha (Ribeiro 1970). These figures point to the high proportion of land rendered economically dead on latifundia and reinforce the observation that it is the frontiers which feed the cities: labour which is consistently denied access to land within the boundaries of the national economy must migrate to the frontier to cultivate its crops. But the massive migration to the frontiers and the labour expended in appropriating the natural environment for the production of agricultural goods represents at the same time a process of creation of economic value. It is only human labour which can create value, and within this pattern of extensive growth it is the labour of the migrants which creates value on the frontiers. This value may be appropriated and transferred to the national economy in different ways. At a later stage in the analysis there will be occasion to examine transfers through extractive activity and transfers through a kind of 'forced levy' practised on the frontier (which will be integrated into a theory of primitive accumulation on the frontier). For the moment attention will be focussed on the progressive integration of the frontiers into the national economy and the transfer which is achieved through the marketing network.

Although the transfer of value from 'backward' to 'advanced' regions in Brazil has been widely debated, especially in relation to the North-East (Furtado 1963; Gunder Frank 1967; Schuh 1969; Fernandes 1968), the transfer from agriculture to industry has attracted relatively little attention, with the possible exception of the transfer implicit in the multiple exchange rates policy of the 1950s (Leff 1968; Bergsman 1970). While it will be important to keep the transfer from sector to sector in mind, the transfer from the frontier is a special instance of the more general process, which depends on the pattern of agricultural growth, i.e. on the very expansion of the frontiers. This expansion has meant continual new production on fresh and fertile soils which has allowed continual downward pressure through the marketing system on prices to the producer which has been instrumental in keeping food prices low: farm gate prices are not related to 'costs' of production (and certainly not to the cost of the declining fertility of the soil) but rather to the rates of return expected from capital investment in commercialisation and industry (Oliveira 1972b).

The transfer of value from frontier to national economy is first revealed in the wide profit margins within the marketing network. The oligopsonistic nature of the network, which is examined below, contributes to force down farm-gate prices, leading, as already observed, to a decline in real prices to the producer over the period 1937–61 (Kahil p. 169). At the same time, however, the ratios of both wholesale and retail prices of agricultural products to the general urban price index tend to rise – at least over the period 1947–63 (Kahil p. 133). Falling prices to the producer and rising prices to the consumer indicate increasing commercial profits. Moreover, from the breakdown of Kahil's figures it is apparent that the ratio of total wholesale prices to urban prices rose much more slowly than that of the wholesale price exclusive of coffee. Producers of coffee, the principal export crop, benefited from a more modern marketing structure where profit margins were established in terms of the final price which could be defended on the world market; whereas profit margins in marketing staples for internal consumption reflect the exigencies of accumulation in the national economy. This reinforces the observation that high commercial profits tend to represent values transferred from the frontier; insofar as these profits are reinvested in industry and not commerce they further represent an indirect transfer of value to industry.

But this is only half the story. Despite the commercial profits, the low and declining prices to the producer assured relatively low prices for staple foods in the cities (and certainly lower than if production had taken place elsewhere but on the frontier). A relatively slow increase in food prices allowed wages to rise more slowly without their real value being diminished proportionately (an interpretation which does not contradict but rather reinforces the more usual 'open market' explanation that low income elasticity of demand for the traditional staples makes it necessary for their production costs to be kept low). In other words, the lower the food prices in the cities, the cheaper it was to reproduce urban labour and the higher the rates of industrial profit and the eventual rate of capital accumulation. And although it may be argued that relatively low food prices may only have achieved a small percentage reduction in the costs of the industrialist, any 'saving' of this sort might make a more than proportional contribution to increasing the rate of capital accumulation given the notoriously low rates of private domestic capital formation in Brazil (Leff 1968). In this way the price mechanism could also promote a direct transfer of value from frontier to industry.

The available data cannot document this direct transfer, but they do suggest that the argument is plausible. Figures from various sources, especially DIEESE (Departamento Intersindical de Estatística e Estudos Socio-econômicos), quoted by Martins (1975), indicate that over the years from 1934 to 1970 an average of about half the wages of a São Paulo working class family was spent on food. Evidently, the lower the prices for food, the lower the cost of reproduction of the labour force.

But the precise relationship between food prices and wages is more difficult to establish – partly because the empirical picture is not uniform. The argument was couched in deliberately relative terms – the relatively low food prices might have allowed nominal wages to rise more slowly without their real value being diminished proportionately. In fact, real wages for many categories of skilled workers fell over the period 1940–62 (Kahil p. 67), while the real wage of the unskilled workers increased by something approaching 50%. The latter benefited most from government minimum wage policy, which became a determining element in relative wage and price levels from the early fifties; real minimum wages increased 37% between 1944 and 1962 in Guanabara (Rio de Janeiro) and 23% in São Paulo (Oliveira 1972b p. 47). On the other hand, as the legal minimum wage rose above their productivity it became increasingly difficult for unskilled workers to find jobs. Overall, it appears that the statement can stand. Singer maintains that average real wages rose by 31%

between 1949 and 1959 (Singer 1972 p. 50) and Kahil argues that the ratio of the wage bill to income in major urban private enterprise rose from 31.9% to 41.0% between 1947 and 1963 (Kahil p. 140). Note, however, that the 'gains' to labour implied here were achieved in the period following the 'redemocratisation' of the country in 1945, when unions were relatively freer, labour laws better applied and when, in particular, working class mobilisation could force large hikes in the minimum wage. The situation altered radically again in 1964; all the gains achieved over twenty years in the minimum wage level, for instance, were wiped out by 1968 (Oliveira 1972b).

Certainly in the conditions of the 'democratic' period, when labour was slowly but surely increasing its participation in domestic income, any 'savings' to the industrialist through relatively low food prices would have appeared significant, and there are suggestions that such savings to the industrialist were substantial. Over the period 1944 to 1965 the general wholesale food price index rose from 22 to 3198, while the corresponding index for the prices of *industrial* products (not the general urban price index) rose from 52 to 5163 (Paiva 1966). This comparison may be slightly misleading insofar as the retail food price indices tend to rise faster than the wholesale over the period, but, nevertheless, Singer feels able to conclude that 'there is no doubt . . . that a good proportion of the profits accumulated by industry has been produced by agriculture and transferred to the industrial entrepreneurs owing to the worsening of the terms of trade between countryside and city' (Singer 1972 p. 19).

Thus in this view the maintenance of the expansive pattern of growth in the countryside contributed to raise the rate of accumulation by furnishing staple foods at low cost (the original prices to the producer being determined only by the cost of subsistence of the frontier peasants). The continuing effect was to reduce the price of urban labour and so allow for higher rates of profit in industry. However, there is no intention of exaggerating the size of the transfer of value in general, or the contribution of the frontiers in particular to Brazil's financial and industrial development. In the first place, the greater part of the appropriation of value in industry was not through absolute but through relative surplus-value: the 31% rise in average real wages over the decade 1949–59, for instance, must be set against the 102% rise in labour productivity (Singer 1972 p. 50) – more than two thirds of which was obviously appropriated by capital. In the second place, it is evident that the value appropriated from any one frontier at any one time will not be significant in terms of the total accumulation within the advanced sectors of the economy. Nevertheless, the argument defends, by definition, the cumulative transfer from many frontiers over at least four decades – and this transfer is itself but one element of a more inclusive process.

The marketing network

As the frontier advances a slowly spreading marketing network accompanies the occupation of the land, integrating the new regions into the national economy. The nature of this network cannot be explained in separation from the system of land tenure which emerges on the frontier. This observation is necessary in the light of analyses (e.g., Smith 1965) which have demonstrated the wide marketing margins and low producer participation in the final prices of agricultural products in Brazil, but which have come to the wrong conclusion that it is the structure of the marketing system alone which brings this about. In defending his thesis, Smith everywhere transparently refers to large farmers (although this is nowhere made explicit) and allows himself to make assumptions which cannot be true for agriculture as a whole. In talking of the frontier, I shall refer the argument to small peasants, not merely because this best reflects the reality of the frontier, but because this reference best demonstrates the nature of the network and its interaction with the system of land tenure.

The multiplication of small peasants within the frontier region (due to the fragmentation of farms observed above) each of whom produce a relatively small surplus, leads to a market situation where supply is 'atomised'. This situation is described by the agronomists as 'micro-production' and by the sociologists as a regime of minifundio. The minifundista who is isolated on his farm is without the means to transport his own produce and without information on market prices (Averburg 1969; CODEPAR 1963). He is obliged to sell to a richer neighbour or to the local middleman or trucker. These may have furnished credit to the peasant earlier in the year and prices for his produce may well have been decided *na fôlha* (that is, in *conta corrente*, where debts and interest are first discounted) long before the harvest is in. Prices are also likely to 'include' transport from the farm and possibly packing. For instance, of the farms under 10 ha in the south-west of Paraná, 70% sell their total surplus to the local middleman or trucker at one time on the farm itself, receiving cash, or payment within 60 days (INDA 1969). The equivalent figures are 62.3% for farms of 10 to 20 ha, 45.2% for 20 to 50 ha and 33% for 50 ha plus. Similarly, the smallest farms receive, predictably, the lowest prices. Where supply is 'atomised' in this way, the commercial centres become very important; they reach out into every *distrito, vila*, *linha*, and *estrada*, and finance, collect, store, transport and export this micro-production. Inevitably with so many suppliers and relatively few buyers the marketing structure which evolves is highly oligopsonistic and exploitative of the small peasant.

Within the frontier region it is the local wholesaler in the primary centre who commands the situation (CIBRAZEM 1967). On the frontier there is usually little industry to provide competition, and, furthermore, while the region remains isolated, poor transport and communications strengthen local oligopsony by raising the threshold of price disparity at which entry into the market becomes profitable. However, these local wholesalers are exploited in their turn by the extra-regional wholesalers who operate their cartels from the industrial centres of the country and who have the advantages of superior capital resources, control of stocks and access to information. These tight-knit groups of wholesalers (principally in the south from São Paulo) capture supply at low prices and transport it to markets where prices are rising continually, making regular 'super-profits'. Thus the marketing system is doubly exploitative of the frontier region as a whole and at both levels the exploitation depends on the dispersion and heterogeneity of supply. It is in this way that prices are manipulated from the top down, guaranteeing returns to the large wholesaler, local wholesaler, local middleman and trucker, in that order.

Credit

The effects of the marketing system on the small producer might be mitigated by an adequate supply of institutional credit but this does not exist. It is true that the number of banks in the country increased rapidly from the middle of the fifties and that official agricultural financing expanded faster than production over the years 1956–65 – progressively more credit being allocated to such cereals as maize, soya beans, beans and rice (19.7% of the total in 1956 and 45.3% in 1965). But against this it should be observed that there are few bank branches in frontier regions (only 2 in the south-west of Paraná in 1963), that most credit is released to finance between-harvest costs and not to finance fixed capital investment (RBG 1970), and that in general there are a series of restrictions on credit to small peasants as a group and non-owners in particular (these are the *posseiros* who are often in the majority on the frontier). In short, it remains doubtful how far increased credit alleviates the position of the small peasant, rather than facilitating the operation of the large enterprises.

The two branches of the Bank of Brazil in the south-west of Paraná are to be found in Pato Branco and Francisco Beltrão: in the area serviced by the first, just 15% of the farmers succeed in getting credit; in the second just 5% (RBG 1970). These figures are reflected at local state level for the value of loans released through the Bank's agricultural portfolio in this region is far inferior to that released elsewhere in Paraná. What is released tends to go to large owners rather than small; in this way the local bank meets the norms of the central portfolio with a smaller number of operations and lower level of risk. And the small peasant who does apply for credit must make several journeys to town and have innumerable forms filled in, without any guarantee of approval. Fares, food and lodging may absorb up to a third of the credit which may anyway meet only a third of his needs. But however bad the bureaucracy there is no alternative: in the south-west these two branches furnished 89% of the value of all agricultural loans within the region in 1967 (ACARPA 1967). Thus the credit available to the small peasant is insufficient and costly – and that is not all.

If the small peasant finally obtains credit, he may live to regret it (Bunker 1978). In the first place it often comes at the wrong time. Requests for credit can only be made in certain periods, depending on its purpose, and delays force the peasant to obtain credit from middlemen while he waits for the bank. In the second place the credit comes without the technical advice which might guarantee an adequate return on it. So the real cost of credit increases and the peasant finds difficulty in paying pressing debts. So he must sell his produce immediately after or even before the harvest at bottom prices which places him back at the beginning of the vicious circle. Credit drawn for fertiliser or selected seeds may finally force the peasant to sell off his oxen or the milk that should go to his family – especially if the harvest is bad or there is illness in the family. Credit may simply increase the rate of exploitation and finally decapitalise his minifundio (Martins 1969b).

Minimum prices

The Federal State operates a minimum price policy designed to increase the bargaining power of the peasant and protect him from violent fluctuations in the prices for farm produce. It is also intended to produce a more even supply of the different staple crops. Prices are fixed by the CFP (Commission for Financing Production) founded in 1943, and raised to the status of semi-autonomous agency or *autarquia* in 1962, but imposed through the branches of the Bank of Brazil – and though legal provision was made for the participation of the private banking sector, this was never allowed in practice. The policy should of course narrow the profit margins available to the local oligopsonies and raise prices paid to the producer, but it is again doubtful that the policy has been uniform in its effects on the producers. Its overall effectiveness has been gauged by the degree to which supply of cereals follows the lead of minimum prices, which is considerable, but it may yet be that only the large farms respond to price stimuli (Pastore 1968). This possibility is reinforced by the measure employed, which is the size of area planted: large farms can expand their production at will, but small farms have neither labour nor land to do so.

The minimum prices for any one product only come into effect at the end of the harvest period, when many small producers will already have sold their produce in order to pay their debts and all the benefit then goes to the middleman. The small producer should however get full minimum price if he exercises the limited option of selling a portion of his produce directly to the Bank. But the Bank discounts the costs of cleaning, sorting, packing, loading and storage – which legally (Estatuto da Terra, Article 85) should be borne by the buyer. Moreover the Bank or whatever buyer will have discounted the ICM (*Impôsto da Circulação de Mercadoria*) which is calculated according to the scale of minimum prices which, for reasons already apparent, are not necessarily available to the small producer. Finally, in addition to its failure to protect the small peasant, the scale of prices is fixed as if the produce were already placed in the consumer market or in the port (f.o.b.), and this provides a bonus for the producing regions nearest the consumer centres and penalises all producers in frontier regions. Add to all this the bureaucratic bottlenecks and lack of bank branches and it is not surprising that weak prices to the producer on the frontier have shown few signs of improvement.

Differentiation of the network

It cannot be denied that this marketing system has suffered modifications in recent years in relation to particular farm products (RBG 1970). To continue with the case of the south-west of Paraná, pigs are now often transported direct from the producer to the state forwarding markets in Ponta Grossa and Guarapuava, or to São Paulo. But here the producer is still ignorant of prices in the major markets and is in no position to bargain. Maize and soya beans are sometimes channelled direct to exporter and industrialist who provide competition for the local wholesalers, but such buyers tend to go to the biggest producers and the effect on prices to the small farmer is still slight. Wheat is brought up by a special department of the Bank of Brazil and the CFP intervenes erratically in the market (as it did in 1967 to buy black beans) (ACARPA 1967), but in both cases the greatest part of the produce passes first through the hands of the local wholesaler who does everything legal and illegal to stay inside the commercial circuits. Their advantage is that they have the warehouses and the agencies do not. In general, government 'protection' may reach export crops like coffee or soya beans, or priority crops like wheat, but little is done to help the producer of black beans, maize and rice.

The only way in which the wholesalers can be cut out of the circuit is through the trucker, who can buy at the farm gate and ship the produce direct to the extra-regional wholesalers or even straight to the retailers in the consuming centres. The trucker usually works with the less 'industrial' crops like beans and maize and can open up the system to a degree by introducing new permutations into the traditional market relationships. He has played an important role in making the market more accessible from the frontier regions and in integrating these regions into the national economy. He is seen by some as embodying salutary tendencies towards a simplification and vertical integration of the marketing network. This is to exaggerate. After all, he operates by profit criteria like other middlemen and changes roles in quixotic fashion: he may sell to retail outlets or direct to the big cities but equally he may sell to the local wholesaler. He is not destined to become the saviour of the small producer.

The impact of the marketing system on agriculture

The principal effect of the marketing structure examined above is to force down prices to the producer, which contributes crucially to the transfer of value from the frontier to the industrial centres. This transfer is but part of a more inclusive process. Indeed it has been shown (Nicholls and Paiva 1965a and b) that in the Vale do Paraíba in São Paulo where some of the most 'technical' agriculture in the country is practised, gross returns may be very high, but after discounting costs of maintenance, depreciation and capital (interest rates) many of the farms actually receive negative net returns. There is a striking failure to distinguish productivity and profit per unit of capital which is nevertheless typical of the primary sector in Brazil where individual enterprises are often 'prevented' from acting in a capitalist manner. The producer is rarely permitted to know the final price for his product, whether bought privately or publicly, and minimum price policy in particular does not respect the agricultural year or the obvious need for estimating costs and returns. So the farmer cannot control his costs (what, how much, when and how to plant) and relate productivity to likely return.

This situation has two important consequences. The first is that, by a peculiar logic, it has been the 'traditional' farm (and particularly that farm on the frontier) which has had the best chance of success in Brazilian agriculture, insofar as it can sell to the monetary economy, but does not need to invest capital for modern inputs. It sells a surplus, this being defined, finally, not as what is not consumed, but as what is produced with surplus factors (that is, land and labour that would not otherwise be put into production for subsistence) (Martins 1971). So the frontiers have continued to expand. The second is that if it is a large owner who is receiving negative returns he will pass them on to the small renter or share-cropper or wage-earner – so that he can still receive a net gain over the agricultural year (if he has inherited or appropriated land or bought it for a nominal sum, any 'money-rent' in whatever form received from peasants on the land will be perceived as 'profit'; if he is a capitalist farmer who has invested in the land and pays wages then he will depress wages or dismiss labour); but the small peasant, the minifundista, has nowhere to pass on his negative returns and so must absorb them directly in a sort of self-spoliation, which I shall call 'decapitalisation' (Martins 1969b).

Conclusion

It may be objected that in the attempt to relate the expansion of the frontier to the development of the national economy the argument has become too deterministic, or at least too 'economicist'. But it is no part of the intention to establish a unilinear causality between the process of industrialisation on the one hand and frontier expansion on the other. Insofar as a determining relation exists between them, it is refracted through contradictions at the economic level, and mediated by the intervention of different legal and institutional structures and practices at the political level. This is material for the coming chapters which debate, among other things, the role of law, the 'inertia' of the bureaucracy, and the informal structure of the State.

It is already clear that there was no formulated policy of a transfer of value from the frontier to the industrial centres of the country; on the contrary, State policies on credit and minimum prices appeared designed to prevent it. But the transfer continued and, as suggested above, State intervention did not merely fail, but appeared to promote what it intended to prevent. Significantly, State intervention had an effect, but not the 'intended' effect. It contributed to shape the specific structure of the market, but did not succeed in slowing the transfer. In specific terms, this intervention in the market (sphere of circulation) did not effect the direction of a transfer determined at the level of accumulation (sphere of production), but did *mediate* that transfer; in general terms therefore the interpretation of the frontier as a response to major economic determinations cannot imply that these simply 'override' piece-meal political attempts to guide its progress. In other words, it is certain that the central connection between frontier and national economy – forged here by the intimacy of commercial and industrial capital – is fundamental to the analysis of the frontier; but this cannot imply that the process of frontier expansion is in any way 'immune' to State policy, monetary or otherwise, or impervious to political and legal intervention. This is demonstrated in the following section on political mediation.

PART 2

Political mediation

4

The legal history of the land on the frontier and the question of dual authority

Traditionally in Brazil land is titled long before it is occupied. Unexplored regions in the interior are likely to have complex legal histories and many areas within these regions will be titled more than once. For a long time settlement of the continent was confined to the coast and the colonisers seemed to baulk at the prospect of conquering the vast and unknown interior. During the colonial administration the territory was conceded in law even before its extent could be judged with any measure of accuracy. This initial legal division of the land (into *sesmarias*) spawned a series of minor concessions. Since that time, for different economic and political motives in different periods, titles have proliferated and as occupation proper has continued to meet with delays and difficulties up till today, few constraints have acted to stem the issue of titles. This titling has created the conditions for conflict over the right to land in nearly all frontier regions.

It is interesting to observe that of the folk figures of the Brazilians the *bandeirante* looms largest in their history because these explorers broke the cultural and geographical boundaries and carried their forays deep into the heartland. But the *bandeirantes* did not occupy the land but merely claimed it for the crown; or for the nation. The actual occupation of the land had to wait for the unsung hero of Brazilian history, the *posseiro*, who was to claim the land by cultivating it. He moves on to the land to work it, so appropriating the environment, and this appropriation has its own tradition. For some 28 years from the time of its independence, Brazil had no land law and what occurred in the interior could only be occupation pure and simple. When the history of Brazilian, as opposed to colonial, land law does begin, it is with full recognition in the first instance, not only of the property rights vested in previous legal claims and titles (deriving from the *sesmarias*), but also of the courage of the small farmer who civilises the interior by making the land produce. Never again would he enjoy such recognition by the State.

In effect, by law 601 of 1850 the Brazilian State asserted full property rights to all unoccupied lands (*terras devolutas*) in the Empire (Westphalen 1968) (and decreed that, in future, land could only be titled if purchased from the State), but recognised previous legal claims, principally those of *sesmaria*, and previous de facto claims (*posse*), as long as the land in such cases was occupied and cultivated by the claimant (*cultura efectiva e morada habitual*). The regulation of the law (1854) made it clear that what was required was a 'consolidated occupation' and not merely a staking out of the land (Zibetti 1969), and for *posse* to constitute a formal right to land it must be registered, surveyed, and finally confirmed by the State. Nevertheless the legal conditions now existed for converting *posse* to private property; at the height of the Empire the State rewarded the *posseiro* and created the incentive for the economic exploitation of the interior. From this time on, *posse* always constitutes a potential property right in Brazil – but one which has required State intervention to make it a reality – and this has occurred only rarely.

At the same time, the sudden appearance of land law in Brazil makes it evident that it is the State which defines what is private property, what is *posse*, what are *terras devolutas* (in 1850 these were untitled, unoccupied land, not destined for public use) – in short, which defines the ground rules for the coming struggle over land. During the Empire however, the State monopolised the land and titled only to those that bought, thus leaving little legal slack for claimants to land to take up. True, it may never have exercised total control over the land (the registration of *posse*, for instance, was carried out very imperfectly by the Church), but at least the line between what was public and what private was clear. The only exception to the legal obligation to purchase land from the State was made in respect of lands on the 'limits of the Empire'. Just as *sesmarias* were conceded by the colonial power, whose priority was to populate the land and not necessarily to produce, so the State in 1850 reserved the right to cede land within 10 leagues of the political boundaries, in order to people those areas for reasons of national security. These areas, with all other *terras devolutas*, belonged to the State throughout the Empire. Come the end of the Empire, and it was only these areas in what came to be known as the *faixa de fronteira* which remained to the Federal State.

With the end of the Empire, by the Constitution of 1891, legal ownership and political control of the *terras devolutas* passed to the local states and hence to the local land-owning oligarchies (Westphalen 1968). From this moment, the legal history of the land, which grows in complexity from this time, becomes also a political history. This complexity is both a cause and a consequence of the key role played by land, both in local politicking and in the grander political history of the country. Legal relations and the process of litigation itself reflect political contradictions and antagonisms which are themselves underpinned by economic determinations. Thus the study of major legal disputes over land as they move through the cumbersome hierarchy of the courts is not merely of intrinsic interest but is essential to an understanding of the dimensions of the political struggle for control of economic resources and, insofar as the land itself is occupied, for the control of the labour process and the appropriation of value.

The crucial change to control of the *terras devolutas* by the local states favoured once again the concession of land to private companies and private capital (Stefannini 1977). It is this circumstance more than any other which marks the beginning of the legal struggle for land in Brazil. By implication it greatly diminished the real possibilities of the *posseiro* achieving property in land and our hero becomes – increasingly over the following decades – the victim of the struggle. Of course, he anyway never appears as a protagonist on the legal stage. The legion of avaricious lawyers and the very length of the legal disputes which 'progress' successively through the pyramid of injustice comprising the different courts, discourage and disqualify even powerful individuals from the tournament. The greater part of the litigation over land is not between individuals but between large economic interest groups and sectors of the public administration (the State in its various bureaucratic manifestations). This is so not only because these are the actors which have the resources to bear the costs of litigation but also because legal judgements may finally be seen to depend on how much political pressure the different litigants can bring to bear on the legal system. Only political power can guarantee the legal control of land and the profit which may accrue from it.

The legal history is also a political history insofar as it is the result of State initiatives over control of land. These initiatives were never uniform or consistent after control of public lands devolved to the local states by the Constitution of 1891. This Constitution left unspecified the question of the *faixa de fronteira* (those areas on the political boundaries). Article 64 reserves areas which are 'indispensable to the defence of the country' to Federal control (along with forts and Federal railways) – but these were never defined. This legal lacuna in the Constitution was to prove central to the subsequent legal history of the land – both as an immediate cause of friction between state and Federal governments, and as a 'precedent' for later Federal attempts to recentralise control over land in Brazil. I shall examine this subsequent history in the context of the three frontier regions in Paraná, Mato Grosso and Pará – before continuing the discussion of the general significance of the legal struggle.

The case of Paraná

As the Federal State did not clearly vindicate its claim, the local state government of Paraná assumed legal right to the lands in the *faixa de fronteira* and titled it accordingly, which it has done continually for 80 years despite repeated warnings and injunctions issued by the Federal government (and judgements handed down by its Supreme Court) which amount to a systematic denial of the state's legal authority in the area. The state could administer the land with the greater impunity the more its own legal system succeeded in creating a separate legal authority (the state system being only loosely articulated with the Federal in an ambivalent subordination). It is this basic conflict, implying a dual authority in the area, which reached contradictory legal decisions and exerted contrary pressures which underpins the area's subsequent legal history. Initially these authorities stayed in the background to censure or sanction the actions of the principal protagonists in the legal drama; their progressive involvement in the plot can be perceived in the succession of scenes leading to the *dénouement* of 1940.

The protagonists are the railway and colonising companies which received grants of land by contract with the state government for the construction of different railways (Foweraker 1974). In general, these land grants were made in lieu of payments in specie, the companies demanding guarantees for their investments. The companies were foreign-based (such as the Brazil Railways Company of the USA and its subsidiary the Southern Brazil Lumber and Colonisation Co., and the Chemins de Fer Sud-Ouest Brésiliens of France), and the willingness of local states and the Federal State to sign away the national territory reflects the difficulty of attracting foreign capital to this sort of enterprise. Difficulties of this order were overcome in São Paulo where the high profitability of coffee made it relatively easy to service and amortise government to government loans, but in other states they continued and the foreign companies (where they did not transfer their rights to national companies and invite participation in order to lower the element of risk) demanded larger and larger chunks of territory as collateral.

It is ironic that in the case of Paraná the major original grant of land had been made by the Imperial State shortly before the demise of the Empire. The state at first refused to recognise the validity of this grant which had been made to a distance of 9 kilometres to either side of the projected track, maintaining that it alone administered the public lands within its borders. Strictly speaking, however, the grant had been made prior to the Constitution of 1891 and, finally, the state seemed to bow gracefully to the overtures of the final recipient of the grant, the São Paulo–Rio Grande railway company (Estrada de Ferro São Paulo–Rio Grande). But by this time (1917) there was no question of granting land along the track (this land lay near the coast and was already settled) so in its stead extensive grants of land were made deep in the interior of the state – squarely within the *faixa de fronteira*.

On further examination it is clear that the state had been a reluctant party to this grant (DTC Paraná 1933) and this can be no surprise when we learn that the total area of land amounted to more than 2 million ha. The state had even begun (with Santa Catarina) to sell land within the area of the original grant to the company but was officially warned against this by the Federal government in 1908. The state remained recalcitrant at this time and suffered several more years Federal pressure before accepting the inevitability of the grant and entering into agreement with the company. It again followed Federal initiative when it then ratified the transfer of most of the rights of the company to another sister company – BRAVIACO (Companhia Brasileira de Viação e Comércio). But it asserted its legal autonomy in another way – by relocating the grants in the *faixa de fronteira* – and in areas already titled by the state or awaiting state approval for the issue of definitive titles. Overall the local state had acted with political restraint in accepting contracts which it had not wanted, but had created a situation whereby private companies, in addition to the state, were now titling within the *faixa de fronteira*.

Following the Revolution of 1930, which was military and nationalist in character (Fausto 1975), the Federal government proved far more sensitive to issues of national security in general and the political status of the *faixa* in particular. One of the first acts of the Federal *interventor* in Paraná was to declare null and void the local state's contracts with the companies and though the companies contested the decree the state won its case repeatedly in state courts and finally in the Federal Supreme Court (1938) (Foweraker 1974). Thus the Revolution of 1930, by decreeing the return of these lands to the state, brought the state and Federal governments closer than they had been at any time since the end of the Empire and, in addition to motives of national security, this can be seen as a reflection of Vargas' policy of conciliation of states' interests during the early 1930s. But with the Federal government coup which brought in the Estado Novo came a radical reversal of policy and a concerted attempt to impose Federal legal rights in the area.

The claims of the Federal government were now unconditional. By the Constitution of 1937 the frontier strip was extended to 150 kilometres; this belonged to the Federal government; local states might grant land within the strip as long as the grants did not exceed 2000 ha, and were reviewed and ratified by the Special Commission of the Frontier Strip (*Comissão Especial de Faixa de Fronteira*, attached to the powerful Supreme National Council – *Conselho Superior Nacional*, and established in 1939). Following these measures aimed at stronger central control, the advent of the Second World War precipitated the coup de grâce, marked by the Federal 'nationalisation' of the companies (São Paulo–Rio Grande, BRAVIACO) – including all lands granted to them in Santa Catarina and Paraná (Westphalen 1968). It was this decree which finally brought local state and Federal State into direct legal confrontation. For the state refused to countenance the sudden descent of the Federal government onto the stage, like some deus ex machina, and asserted its legal right to the lands by virtue of the decree of the Federal *interventor* in 1930 and the decision of the Supreme Court in 1938 (DTC Paraná 1948).

Dual authority

In this way two constituted but contrary authorities came to be present in the west of Paraná from the beginning of its occupation – the local state and the Federal State. Both tried to vindicate their rights to the area with varying degrees of success at different times, and both accused the other of prejudicing the region's development. The Federal State finally took the view that the local state is wrong from beginning to end: it titled the land in the strip which is Federal property and, besides this, persistently ignored all due legal procedure (such as surveying the land before titling; such as issuing title only if the applicant begins to work the land within 12 months). In fact, state titling was so disordered that titles were frequently issued on top of others issued by the state itself (Costa de Albuquerque 1970). The state replied that Federal claims were very hypothetical and its legal arguments of relatively recent origin. Only after 1945 was a more limited strip of 66 kilometres seriously defended and even then, the Federal State did not match words with action by proceeding to a legal definition of the area. On the contrary, it deliberately avoided the property problem by leasing land while reserving final property rights to itself (*Relatórios* DGTC Paraná n.d.). Federal leases disregarded state titles in the area so that it became increasingly possible for the Federal State to lease land already owned by a state title-holder. Not surprisingly this led to conflicts between leaseholders and title-holders with sometimes fatal results, so generating a permanent social tension in the area. Moreover given the dual authority within this area every local conflict carried the further dimension of a Federal–state confrontation which could only be resolved in the Supeme Court.

What we must now trace is the course of the confrontations between different sectors of the State which are contained by the institutional structure of the courts and modified according to the norms of legal procedure in Brazil. These confrontations continue and increase as the land takes on value and private interests fight for control. Railway companies, so-called colonisers, and private speculators – often belonging to large economic consortia and often enjoying high level political connections in the local state administration – are now actively seeking legal control of the land. With sufficient political power to defend their title they can then use its legal sanctions to promote their economic interests. With continuing disputes over titles at every level and the pattern of private property ill-defined, the ownership of the land may change not only by commercial contract but by the decisions of the courts. This is the way that economic interest is 'represented' on the frontier: elaborate legal statements may clothe simple speculatory motives; a successful appeal may bring windfall profits to a colonising company. But while it is true that control over the region's resources is disputed between different economic groups and that the complicity of the local state is necessary in securing the titles which legitimate the control, remember that prima facie the principal dispute is with the Federal State which claims these lands in toto in the interests of national security, and Federal decrees invalidate the legitimising titles of the state. Finally then, grounds for conflict emerge between the economic groups and capitals on the frontier and the agencies of the Federal State in Paraná. The outcome of the conflicts depends, in the first instance, on the legal system, and this in turn proves sensitive to its political context.

The fifties: a period of state ascendency

At the beginning of the fifties the situation had evolved where both Federal and state governments were issuing title or lease to land; individual companies had also titled 'within' their respective areas (and not always obeying even this legal imperative). All local conflicts engendered in this way carried the seeds of conflict between state and Federal State.

The state now took the initiative in attempting to annul all titles issued by private companies and especially BRAVIACO and the São Paulo–Rio Grande within the various judicial districts of the west, and its control over its own legal apparatus ensured that writs to this effect were issued to the judges of the region. For its part, the Federal State had placed all 'nationalised' lands in the legal care of SEIPU, the bureaucratic agency which was uniquely responsible in Federal eyes for the sale and titling of the lands. SEIPU issued an injunction to prevent the titles being cancelled, so inaugurating a long series of legal manoeuvres which finally bound both parties in legal stalemate. Not until 1962 did the state finally succeed in cancelling these titles (DGTC 1966).

This is one skirmish amongst many, but it illustrates the broad lines of interaction in the fifties. The state laid legal siege to all Federal positions, while the Federal government had to rely on its agencies in Paraná to achieve the control to which it aspired – and they were largely ineffectual. Following SEIPU came INIC, the first of a series of agencies designed specifically to deal with the land problem (INIC, SUPRA, INDA, IBRA, INCRA) and in 1958 INIC assumed responsibility for most of the land in SEIPU's care (Procuradoria Geral 1968). But like its predecessor INIC became involved in fruitless legal wrangles, and the continuing impartiality of the Federal courts worked to its disadvantage. In short, the Federal government failed in its intentions. This failure, it will be seen in the following chapter, effectively left the way clear for intense independent activity by private capitals and economic groups – especially those allied with the local state administration – which moved into the front line of the battle for land.

The local state's freedom of action depended on the configuration of Federal politics over the period, and, in particular, on the political influence of the PSD and its leader Moisés Lupion, governor of the state from 1946–50 and 1956–60, and one of the great machine politicians. In his earlier term he had been one of the pillars of the Dutra administration and continued influential before 1955. As mentioned in Chapter 1, he himself headed an economic group with important interests in the west, not least of which was the title to Missões and part of Chopim (together forming a huge estate of 425,731 ha) which Lupion's firm, CITLA, had obtained from SEIPU in 1951. This transaction was blatantly illegal and unconstitutional involving extensive bribery, nepotism and corruption (Mader 1957), but indicative of Lupion's influence is the marked reluctance of Vargas to instigate a Federal investigation despite the insistence of the Chancellor, Oswaldo Aranha. Any such investigation would have weakened the PSD faction in Paraná. In fact, after renewed pressure and protest Vargas agreed to a Federal investigation, but his suicide followed shortly afterwards, and Lupion was saved by the bullet.

Just how typical was the weakness of the Federal government apparatus in response to pressures from the state political machines is not a question to pursue here, but it is clear that the PSD machines were highly autonomous organisations in most states, and never more so than in Paraná in 1956, when Lupion was returned to power at the head of a PSD machine that was essential to the political stability of the Federal government. Kubitshek had been elected on a PSD ticket but had only polled just over one third of the votes cast, and it had needed a 'preventative' coup to guarantee his accession. Therefore, Kubitshek could not afford to alienate the Paraná PSD if he was to build a viable political base, and Lupion was secure enough to allow CITLA to begin massive and violent 'colonising' operations on its land in the west of the state, immediately creating widespread social unrest. All PTB and UDN members, and fully half of the PSD (23 out of 45), in the legislative assembly telegrammed the President demanding action (*Diario da Tarde* 1957). Kubitshek promised this in the Federal Senate in late January 1957, but nothing came of this; for when he sent a dispatch to the National Supreme Council, the PSD in Paraná threatened to secede from the national party, and this was sufficient to intimidate Kubitshek at least for a time (Lacerda 1957). On the other hand, a petition from just 5 prefects demanding the closure of CANGO, the Federal colonisation scheme in the CITLA area, was attended to immediately. By cutting Federal funds to CANGO Kubitshek ensured that it would not provide effective opposition to CITLA's operations in the area.

This example illustrates how the state administration, more deeply involved than ever in economic speculation in land, enjoyed almost complete autonomy to use and abuse its power of political patronage and exert unprecedented pressure on the legal apparatus. While the state and Federal agencies remained engaged in complex legal debates over ownership, Lupion continued to title in a way which became reckless after 1958 (GETSOP 1966). He titled on top of estates which already had title; he titled to buy political support and pay political debts. He would survey and title land to applicants who in reality did not exist, and then these phantoms gave power of attorney to economic interest groups allied to the state administration. If any other title-holder complained, then police and local authorities would be expected to uphold the new title. These 'false petitions' (*procurações falsas*) were just one of the mechanisms which complicated and confused the legal situation – reaping multiple and violent repercussions in the region.

After Lupion's departure from office the gravity of the social situation in the area of CITLA's 'colonisation' prompted the Federal authorities to forgo for once the luxury of a protracted legal debate, and in 1961 they expropriated the area 'in the national interest'. Just one year later the Federal government decreed the establishment of GETSOP (Grupo Executivo das Terras do Sudoeste do Paraná), a special agency designed to execute the expropriation and enter into agreements with the local state in order to bring peace to the area. GETSOP had heavy Federal backing and had no difficulty in persuading the state under governor Ney Braga to sign an agreement on joint action which ushered in the Grupo Mixto Federal–Estado do Paraná. A new era of legal and administrative cooperation had begun on Federal initiative.

The sixties: the federal reaction

The sixties saw a steady increase in Federal pressure on the local state administration especially after the Revolution of 1964. This pressure continued to be exerted on and through the legal system, for this was still the most direct channel for achieving cooperation through confrontation, and the Federal government now began to get the 'right' decisions from the courts – even decisions which ran contrary to precedent. It may still have been hard to improve the performance of the land agencies, but at least the Federal State now found it easier to direct its own Supreme Court.

Evidence for this is found in the judgements handed down on the principal case pending in the court at the time of the Revolution, that of SEIPU and the state of Paraná. The state was defending its right to annul the titles issued by the private companies (São Paulo–Rio Grande and BRAVIACO) while SEIPU argued that these lands had anyway been 'nationalised' in 1940 and the state therefore could not meddle. The court judged in favour of SEIPU in 1965, a decision which ran contrary to a precedent created only three years earlier in 1962, and which clearly reflects Federal political pressure (Procuradoria Geral 1968). However, this major legal success in the Supreme Court soon proved a pyrrhic victory. The court orders issued to implement the ruling were never carried out, principally because of pressure brought to bear on the Federal Attorney's office. It is obvious that powerful interest groups were being brought into play, and mobilising their forces to block implementation of the orders, and protect the thousands of state titles issued in the west since the state's first attempts in 1951 to annul the titles of the private companies. An indication of the weight of these economic interests may be gained from the number of titles in the west today, which total some

300,000, deriving from 24,500 original state titles (Costa de Albuquerque 1970).

Despite difficulties of this order the Federal government has pursued its policy of getting the right decisions from the courts and of forcing compliance through pressure on the courts. There were other celebrated cases too numerous to mention. But despite the 'victories' it is perhaps too evident that the Federal State succeeded in controling neither the legal definition of the region nor its settlement. But it was obliged to go to law nonetheless in the face of the legal conflicts and the resulting real conflicts and violence in the region itself. The nature of this obligation is especially illustrative of the ambiguities recurrent in the relationship between Federal State and local state. If, in search of a legal solution the local state annuls its titles, it must pay compensation, so it fights every case in the courts; similarly wherever the Federal State expropriates it must pay compensation. So it fights every case in the courts. Evidently it is in the interests of the local state administration and local economic interests to advocate expropriation, because with expropriation the state title is not technically annulled so the Federal government pays, and not the state. But as INCRA – on behalf of the Federal government – usually expropriates only *por interêsse social*, that is, in cases of extreme social tension, it is in the interests of the state and private capitals to promote social tension and revolt on the frontier.

The case of Mato Grosso

The legal history of Mato Grosso reveals the same order of political contradictions, but the emerging conflicts are expressed differently. In general, the case is less complex, and this is partly the result of the reduced role of the local state in this history. In contrast to Paraná, one single company, not several, monopolised the land in the south of the state, and for many years dictated to the state government; moreover, early Federal intervention in the form of direct investment in the North-West Railroad forestalled conflicting land grants to competing companies. In this way, the lines of confrontation drawn between Federal State and the company tended to exclude the local state; but the confrontation exacerbated, in turn, a regionally based conflict within the state, which was to sharpen with the advance of the frontier into the south. The Federal intervention can be seen to foreshadow similar initiatives in Amazônia, rightly dispelling any illusion that these are without precedent. In this sense the case of Mato Grosso 'bridges' that of Paraná, where the local state is centre stage, and that of Pará, where the Federal State is clearly in control.

Several years after Brazil's war with Paraguay an expedition was sent to Maracaju in the south of Mato Grosso with orders to fix the new international boundaries agreed by treaty. In the party was one Tomas Laranjeiras, a Portuguese who had been resident in Buenos Aires. He saw the economic possibilities of the native maté groves spread throughout the south and with the constitution of 1891 came his opportunity to realise them. In 1892 the state government granted a lease on the land to Tomas that he might exploit the maté in the public lands along the border with Paraguay and just one year later the Maté Laranjeiras Company was founded which was to occupy the region (Correa Filho 1957). In effect it occupied vast stretches of land and later destroyed huge areas of forest to make way for the maté trees which were now planted by the company. The Indians within the area were killed and labour imported from Paraguay; the company used its own police to keep out Brazilian colonists who might pose a threat to its monopoly.

Like the companies in Paraná and Santa Catarina the Laranjeiras was foreign-based – an Argentinian company backed by British capital. Already in 1907 it was mobilising political support within the state behind its petition to extend the area of its monopoly and to found successor firms which would work to stem the flow of colonists who, in the words of Miguel Murtinho (brother of the state chancellor, Joacquim Murtinho and spokesman for the company) threatened to become a 'state within a state'. The company was also demanding the right to buy the land of its choice and not merely lease it (it was in a commanding position to do so having bought up the paper assets of the Banco do Rio Branco and Mato Grosso when the bank failed). Despite the agitation of the Murtinhos the legislative assembly rejected this petition only to accept it five years later when a new *situacionismo* had been created by the immense powers of bribery and vote-buying of the company. Those opposed to the petition including the governor, Celestino Corrêa da Costa, manoeuvred to limit the area of land under the company's control which was fixed in 1915 at 1,440,000 ha (the area proposed by the company has been four times this) and successfully prevented the lease being converted to title. In fact, the only 'state within a state' was the company itself which was economically more powerful than the host state of Mato Grosso. The governor in 1924 was again Celestino and his message to the assembly noted that the annual budget of the state (5000 contos) compared ill with that of the company (30,000 contos or more). The company used this economic power and its police power to stop the northwards movement of colonists (contemporary reports indicate that there may have been as many as 20,000 migrants in Ponta Porã, the extreme south of the state), and maintain its monopoly over the maté groves.

Maté Laranjeiras dominated the state politically, and by its contract with the state dominated the region of the south. This contract was progressively limited to 200,000 ha or so, but the company continued to occupy at least 600,000 ha, and some say 1,000,000 ha. Things looked set for decades. But this was to count without the interference of the Federal government. The first Federal initiative was taken in ignorance of the company, but was anyway to change the whole political economy of the south. It was following the menace of war with Argentina in 1907 that the Barão do Rio Branco began to build the North-West Railroad (Estrada de Ferro do Noroeste). Belgian and French firms carried the line as far as the border with Mato Grosso (in return for the usual concessions of land) but no private company would extend the railway beyond the border for lack of commercial interest. So the Federal government itself put up capital for the project in the interests of national security. This piece of 'political infrastructure' not only opened up the south economically, transforming small towns into important commercial centres, but greatly simplified the subsequent legal history of land in the region by pre-empting the usually litigious grants of land to private railway companies. The huge differences between the frontier experiences of the west of Paraná and the south of Mato Grosso (both of them areas within the frontier strip) are in no small part attributable to this early example of Federal intervention in this region.

Unlike the first, the second Federal initiative was directly concerned to counteract the political power of the Laranjeiras company. In 1943 Vargas founded the Federal Territory of Ponta Porã and in the following year annulled the rights of the company within the Territory. He rescinded the contract in order to break what was patently private and foreign control of a region of the political frontier (Castello Branco 1951). The intention behind this Territory as with others founded in the same year (such as the Federal Territory of Iguaçu in Paraná) was to open up and develop these frontier regions through colonisation – in the interests of national security (Esterci 1972). As the local state governments had insufficient resources to colonise the Federal government must do the job, and the Federal government was sure of its legal right to do so – both territories lying within the frontier strip. In Paraná little progress was made because of the Territory's isolation; in Mato Grosso the story was much the same overall with the significant exception of the Federal colony at Dourados (Porto Tavares 1972). In Paraná the Territory impinged only slightly on the complexity of the region's legal history; in Mato

Grosso the principal result of the Federal intervention was to liberate those lands leased for so long to Maté Laranjeiras and so herald the rush for land in the south.

Up till this point in time several developments distinguished the legal history of the south of Mato Grosso from that of the west of Paraná and made it more peaceful and less subject to conflict overall. One reason for the relative absence of conflict was precisely the monopoly of Maté Laranjeiras and the company's police force which kept colonists out (admittedly by the use of violence); another was the relatively slow pace of settlement and, as already noted, the absence of litigious land grants to railway companies; while yet another was the sheer quantity of land. Land being abundant, it was cheap, and so there was little motive to fight over it (no-one suspected it would increase in value so fast). These same arguments apply with almost equal vigour to another potential source of conflict, that between state and Federal government over the land in the frontier strip. As elsewhere the Federal State asserted its ownership of these lands and issued long-term leases to settlers and 'landowners' within the strip. The state government had abundant land to title outside the strip and so did not quibble. There were one or two disagreements when the state did title in the strip (as at Bodequena, where the Federal State refused to recognise state titles) and in the sixties in areas of intense migration such as Iguatemí the Federal State has expropriated to avoid conflict (Molina 1970; INCRA 1970). But in general the potential conflict arising out of a clash of interests never took place.

While this is true, the Federal Territory did provoke a conflict of a quite different order, but similarly related to the advance of the frontier. This has to do with the consciousness in the south that its potential has been crushed politically in Cuiabá, the capital in the north. As early as 1896 there existed a movement to divide north from south led by landowners and lawyers which, while it met with no success, clearly demonstrated the depth of resentment of southern landowners at their exclusion from the political life of the state – all political posts being filled by *cuiabanos*. Throughout the First Republic the fear in Cuiabá that development in the south might wrest political leadership from the northern politicians led to strong pressures to allocate resources away from the south. Requests for schools and roads were systematically refused. Animosity created in the south in this way was strengthened by the belief that large states are necessarily backward (and the example of Paraná seceding from São Paulo in 1850 was invoked). In 1930 Vargas promised to study the problem were the Revolution successful, but nothing emerged from this success until the south joined the *paulista* revolt of 1932, on condition that victory would make the south 'independent'. In the event the future 'governors' of the south were forced to flee to Paraguay. Nothing daunted, agitation continued, and the south welcomed the Federal Territory; despite the ignominy of direct Federal control, it did give more expression to the south as a separate region.

In the light of this history it is evident that the movement to rescind the Territory in 1946 (upon the 'redemocratisation' of the country) met with a mixed response in Mato Grosso. Politicians in Paraná and Santa Catarina were already fighting to re-integrate the lands of the Federal Territory of Iguaçu into their states' jurisdiction, and João Ponce (interventor during the post-1930 period and Federal deputy for Mato Grosso) introduced a clause which would bring the same results in his state. João Ponce, needless to say, was *cuiabano*, as was the Federal President, General Dutra, and this happy coincidence guaranteed all the Federal support he required. Ponce was obviously seeking to preserve the integrity of the state, but other interests were working to create a new local state Territory on the dissolution of the Federal one, which would have its centre in Campo Grande. This movement was certainly separatist in its implications, and strongly opposed to the conservative phalanx led by Ponce. Given the arraignment of forces at Federal level it was bound to fail, but the rivalry and resentment it embodied must be remembered in considering the subsequent legal history of the land as manifested in state initiatives on land settlement (see Chapter 5). For the lands within the Territory were important electorally in Mato Grosso and the further settlement on the frontier proceeded, the further the political balance would swing to the south and the greater electoral representation the south would achieve. Settlement of the land after 1930 is politically advantageous to the south and threatens the livelihood of the northern politicians who work to sabotage it. But of course settlement is also a political issue in a broader sense; politicians from north and south applied for land in order to speculate – as elsewhere in Brazil – and most political pressures were exerted not about regional rivalries but in order to achieve profits.

The case of Pará

In comparison with Paraná and Mato Grosso the legal history of Pará is only remarkable for how few contradictions of any significance emerge in the first 70 years following the first Federal Constitution. This is not to say that the legal administration of the state was any more effective than elsewhere, nor that the history itself is not intrinsically complex. The state had inherited a huge number of registered *posses* from the Empire (22,611) but only very few of them had been 'authenticated' (639) (ITERPA 1976), a situation which would, as always, provide fuel for conflict on the ground at some point in the future. In response to this and other difficulties it encountered the state revised its land law at least four times, even before 1930 (ITERPA 1976). But in its broad lines the legal history over this period was uncontentious, continuing to be defined by the principles and substance of law 601 of 1850 and the particularities of the law itself reflect the predominantly extractive nature of the economy (first rubber, then cashew nuts).

In 1930, with the rubber economy in decline, and cattle confined to the Ilha do Marajó and small areas of the Baixo Amazonas, there was virtually no interest in the public lands of the state and the only issue to emerge by this time was the active dispute over the leasing of the cashew nut groves: the value of the groves rose rapidly as cashew nuts found acceptance and high and stable prices in the international markets. But even here no major conflicts occurred, least of all between Federal and state governments. The Federal interventor in Pará in the 1930s, Magalhães Barata, cancelled all state leases, which returned to the state for further study, and this suited the state's interests very well. It was Barata too who decreed the new corpus of land legislation in 1933, and its 256 articles remained in force until 1966 (ITERPA 1976). The only incident worthy of mention in the intervening period is governor Augusto Corrêa's initiative in 1954 in imposing a 100 ha upper limit on titles to agricultural land, a measure which many have corresponded to the economic reality of the small farms along the Bragança railway, but which proved totally inadequate for the needs of the rest of the territory and the exploitation of rubber, cacao and cashew nuts. As these demanded far greater areas, literally hundreds of titles were then issued by the state in plain knowledge that they were illegal. Some of these titles were later revised, but some remained as mines in the path of a clear legal definition of the territory.

It was in 1959 that this rather sleepy legal scene suffered the impact of the arrival of the Belém–Brasília road, and it is here that the contemporary phase of the legal history of the state begins. The arrival of the road opened up the huge and fertile region to the south of Belém, causing rapid development, a vertiginous rise in land prices and a profusion of land problems. At this time the Department of Works and Land of the state possessed neither a survey map of the state territory, nor even a collection of extant land legislation, let alone an administrative structure capable of meeting the challenge posed by the impact of the road (Monteiro B. 1962). In its wake, the road brought an influx of land grabbers and speculators, and vast areas of the huge expanse of public lands within the state were sold in conditions of near anarchy. Even if land had been sold according to the law there was no possibility of verifying the scope of claims (the Land Department disposed of only one jeep to transport its personnel), and there followed the usual legal deficiencies, fraudulent surveys, superimposition of titles and phantom applications for titles. In short, the state's precarious legal structure collapsed under the impact, but only having previously been undermined by state politicians and bureaucrats. For even before the completion of the road the Land Department received a special batch of 200 titles for signature, each of thousands of hectares to either side of the length of the road. The titles already carried the signature of the governor and awaited that of the Minister of the Department, Benedito Monteiro. It is true that a lot of land had been sold in the state during Monteiro's tenure, but, in this case at least he nurtured 'romantic' notions of a colonisation by small farmers along the roadside and withheld not only his signature but the titles themselves. This evidently hurt, if only temporarily, powerful economic and probably military interests: come the Revolution, Benedito's romanticism earned him 6 months in gaol, 60 days incommunicado and one of the first *cassações* in Pará.

With the Revolution of 1964 we witness the second major impact on the legal history of Pará which is the spate of Federal legislation which follows it. Local state legislation now begins to respond to Federal initiatives in order to match the spirit and letter of Federal law. In 1964 the Federal government drew up the Estatuto da Terra, its major piece of legislation for reforming the occupation and use of the land in Brazil (Messias Junquiera 1964); and in 1966 the state replied with a law which made full titles conditional on the effective occupation and exploitation of the land (a condition that was imposed more than a century earlier for the conversion of *posse* to property in land). Then in 1966 the Federal government created the Legal Amazônia, defining the area of Amazônia which can benefit potentially from a whole range of Federal development programmes, and fiscal and credit incentives to private capital to enter the area, and in 1969 the state replied with a law which greatly increased the budget of its agricultural ministry and hence the Land Department, and fortified the Agrarian Development Fund of the state. In the context of the uneventful legal history of the state these were important changes, but they pale into insignificance beside the import of the new Federal Constitution of 1969 which by its Article 4 defined as Federal property all public lands 'considered indispensable to national security'. From this moment on the legal history of land in Pará was subject to the key decision of the Federal State which would define the scale of Federal pretensions and determine the political future of Pará. The decision came in 1971 by Decree-Law 1164 (ITERPA 1976). The Federal State through its land agency INCRA would directly control all land to 100 kms on either side of all roads constructed, under construction or projected within the area of Legal Amazônia; at the same time it would expropriate 64,000 square kms in what is known as the Altamira polygon in the state of Pará (the biggest expropriation of land ever effected by a capitalist State). The law was altered 3 times in the following 5 years but on each occasion only to increase the area under Federal control. At a stroke, the state of Pará had lost control of 80% of its public lands.

If we include those lands in the frontier strip of 150 kms (some 150,000,000 ha) it is now the case that INCRA – the last in the long series of Federal land agencies – controls 50% of Legal Amazônia, and 30% of the national territory, or some 311,000,000 ha (Cardoso and Muller 1977). It is without doubt the largest landowner in Latin America and, very probably, the capitalist world. In political and economic terms, the lands it controls in Amazônia and, more specifically, Pará compose the new frontier strip (*faixa de fronteira*). Of course, these lands do not exist on the political boundaries of the nation, nor is there any connection in law between the strip of ten leagues of 1850 and the strip of 100 kms to either side of the road network in 1971. But the political content of the two is very similar. Both were instituted in the interests of national security and colonisation – or 'development'. National security in 1850 implied peopling the political frontier in order to prevent the incursions of foreign powers; national security in 1971 implied the control of the advance of the economic frontier in order to avoid conflicts over land – which, for the Federal State, in turn carry the seeds of 'subversion'. The new powers invested in INCRA fitted the Federal State's plan for an integrated development of Amazônia. Following Medici's visit to the drought areas of the North-East in 1970 this plan included opening up and settling the vast Amazon region with labour ('surplus population') from the North-East (Muller et al. 1975). INCRA was to execute official colonisation, promote private colonisation and carry through the agrarian reform (proposed in the Estatuto da Terra but still existing only on paper). In order to prevent a disordered occupation – as had occurred so frequently within the original frontier strip – the land was taken from the local state where, in the case of Pará, the Federal State alleged lack of administrative structure and competence to deal with the problem; and the Federal State took responsibility for directing what had in Brazil's history always been a largely spontaneous advance of the economic frontier. This it hoped to do by maintaining a massive military and bureaucratic presence in the Amazon region.

The state of Pará did not contest the measures which left it in control of a mere 30% of the total area of the state. In principle, the Federal initiative was constitutional, and it would only require a further decree for the Federal State to take control of all public lands in the nation. If this comes about, then the Federal State will have restored in toto the status quo ante, or, in other words, it will have revoked Article 64 of the 1891 Constitution. Despite the constitutionality of the changes, in reality the Federal State has used its 'discretionary powers' to take over the land much as Vargas did in setting up the Federal Territories in 1943; states' interests see the decrees not as discretionary but as arbitrary. Simultaneously with the creation of this new 'strip', the Federal State was solving the problems of the old. In 1966 new legislation appeared which allowed the Federal State to ratify irregular local state titles and concessions in the frontier strip (Min. Agricultura 1968), and in 1975 the Federal State issued a new law confirming that all states' titles in the strip to areas of less than 10,000 ha would be ratified (*Estado São Paulo* 1975). The law which has the stamp of approval of the Supreme National Council, specifically embraces all land within the various frontier strips (of 66, 100 and 150 kms) claimed in different periods by the Federal government.

The local state could do little in the face of this massive Federal intervention into areas of former state competence. What it did attempt to do was put its own house in order. In April of 1975 all land sales were frozen by governor Alysio Chaves and 6 months later the state set up its own land agency ITERPA (Instituto de Terras do Pará) (ITERPA 1976). ITERPA replaced the former Land Departments of the Ministry of Agriculture, which had proved ineffectual in dealing with the thousands of law-suits and petitions for land spawned by the arrival of the Belém–Brasília road, and the spurt of development (principally in the form of cattle-raising 'projects' largely financed by SUDAM's fiscal incentive schemes). A partially state-owned land company, COTERCO, had also been tried, but worked too slowly, and without any clear idea of where to work, for it to be successful. ITERPA is the executive organ of the state in everything that refers to the public lands of the state and its brief is to collaborate with all Federal agencies including BASA (Bank of Amazônia), SUDAM (the Superintendency of Amazônia) and principally INCRA, towards solving the land problems of the state. ITERPA faces problems ranging from *posse* and violence to fraud and law-suits; it is presently engaged in processing approximately 10,000 petitions for land while at the same time it intends to change to the sale of land by auction, not petition. It is hampered by lack of resources and poorly articulated administrative cadres. But of all the problems it faces by far the most serious and one which necessarily compounds its legal and administrative difficulties is the task of defining with INCRA the precise geographical limits of the new 'strip' of Amazônia.

INCRA and ITERPA are both land agencies working within Pará and neither know the limits of their respective authority over the region. As INCRA has not yet surveyed all the lands of its new 'strip', obviously no maps exist which can instruct them in the divisions of their areas of competence. Even if INCRA were assured of these divisions this is no guarantee that ITERPA would accept them. In a recent court action ITERPA fought to recover certain lands which were expropriated by INCRA within the Altamira polygon, and was successful; however, of these lands some fall within the strip of 100 kms to either side of the principal Amazonian highway, the Transamazônica, which further complicates the situation. And the question is not merely one of geographical divisions. Corresponding to the different areas of competence are two separate legal systems, and as soon as the Federal State even proposes the construction of a road then immediately confusion arises over whether tens and hundreds of individual *posses* and properties fall within the Federal or the local state's judicial districts. This only compounds the confusion already existing over which geographical area falls within which state judicial district (geographical proximity not being a safe indication). Finally, while INCRA's 'strip' includes roads at the planning stage, the land only legally becomes Federal once the budget for the road is approved, and in the interim period ITERPA is responsible in principle for controlling the rush for land and the proliferation of *posses* which news of the road provokes (as happened on the projected road to São Felix do Araguaia). Thus the conclusion is inescapable that once again there exists in these frontier regions a situation of dual authority, where the limits of authority are not clearly defined, and where the objectives of the two authorities may clearly diverge.

Of the two agencies INCRA has the advantages of far greater administrative and financial resources and a coherent and well-publicised set of objectives. These objectives, which include ideas of development and productivity, are imposed by the Federal State, and after 1973 have been orientated principally to the implantation or promotion of the large enterprise as the proper vehicle for the occupation of the Amazon region. ITERPA, in comparison, works with exiguous resources and the far narrower objectives of defining the legal rights to land. Prima facie INCRA will favour big capital while ITERPA may occasionally be open to pressures from smallholders and *posseiros* – and indeed is required by the state constitution to authenticate, free of charge, all *posses* of less than 100 ha (which it rarely does for lack of resources). In so far as either is corrupt, INCRA may only be corruptible by big capital, while ITERPA may be slightly more democratic in its corruption. Similarly, being a state agency with a legal brief, ITERPA is part of the state's legal apparatus and so takes a different attitude to INCRA to the thousands of titles issued by the state since 1891; these titles often conflict with the more recently acquired titles of the large enterprises which INCRA is concerned to ratify (recent Expositions of Motives, nos. 5 and 6, of the National Supreme Council, look to resolve all legal questions in favour of the large enterprise and its economic 'projects'). Finally, its very lack of resources forces ITERPA to sell land as fast as possible, and this highly commercial attitude to land can conflict with the developmental plans of INCRA. ITERPA is now selling yet faster in the face of the recently mooted possibility that INCRA will take over all public lands in the state. The suppression of ITERPA would be a logical last step and the shortest way to dismantle the recently reconstituted dual authority in the region.

Legal conflicts and political negotiation on the frontier

It is evident from an examination of the legal history of the land on the frontier that the question of legal rights to the land nearly always promotes conflict. Having looked at these conflicts at the level of the legal and political systems, it is evident that disputed rights to land often derive from different sources of title to land. Titles are issued by private companies, by the local state and by the Federal State. At this level of examination the distinction between public and private is not very useful: both public and private bodies may seek to control land in order to speculate or realise profits of some sort. Private capital and companies may seek the complicity of the local state government (as in Paraná and Mato Grosso), or of the Federal government (as in Pará today); but as title to land always originates somewhere in the State (in the public realm), then where these titles conflict the legal history develops across major political contradictions, which can provoke direct confrontation between different sectors of the State.

The occupation of land is an economic process before it is a political one, and it is economic motives which bring the different capitals and economic groups to the frontier. Then, however, the problem of gaining control of land becomes political and these capitals and groups seek title to the land which can legitimise their economic activity. But decrees and titles from one 'source' may invalidate the titles of another. In other words, where they seek title in an area of dual authority the political contradictions surface immediately. Usually the major contradiction lies between local and Federal governments but, as in the case of Mato Grosso, it may lie between different factions of the local state. In this way competition for property in land creates legal debate and confrontation which can only be understood politically.

The original disputes are often between private capitals and companies which are also title holders. But in ways which are now clear these disputes are transformed within the legal system into disputes between sectors of the State. Ironically, with the local states and Federal agencies embroiled in legal wrangles, the economic groups and capitals may enjoy greater freedom to speculate in land, and exploit land and labour on the frontier. They have greater licence for their economic operations precisely because the legal situation is so ill-defined (contrary to the institutionalised relationships of the larger society where the legal relation of property is essential to economic activity and accumulation). This economic activity centred on speculation in land is examined in the next chapter, which shows how the conflicts deriving from dual authority are reproduced in myriad and violent form between *posseiros* and peasants on the ground, and title-holders, lumber and colonising companies, and their hired gunmen.

Evidently the legal system is used and abused in the prolonged struggle for the control of land, and the pressures exerted on the legal system both by private capitals and sectors of the State make it a partial instrument of political control rather than an impartial instrument of justice. However, the study of the legal history allows us to perceive much more than the occasional corruption of the courts. What is important politically is the degree of autonomy the legal system maintains while absorbing pressures from different directions. In this way the system provides channels for bringing together different groups, public and private, in confrontation. In this sense, the law acts as a mechanism for bringing different sectors of State and private and public institutions into the same political arena, and provides a structured context of communication within which contrary demands can be presented in the same language. In other words, the legal system provides for interest representation, and it may be that only after these interests have been expressed in terms of legal conflict can the bureaucracy respond to them. Within the context of an authoritarian regime with limited pluralism and co-optation these legal procedures may be seen increasingly as 'an adequate equivalent of more collective, political expressions of interest conflict' (Linz 1964). Almost paradoxically the legal debate which reflects and spawns so many conflicts is also the language of negotiation. What we witness here is authoritarian dialogue.

It was asserted earlier that this legal history is also a political history, if only because law is made by the State. The various laws and decrees and constitutions represent political initiatives, and the legal debate over land is of course political activity (later it will emerge that the intricacies of the law-suits reflect the complexity of the play of interests within the bureaucracy). Evidently it is not the case that the range of political activity and variety of political decisions referred to here are directly determined by economic or 'structural' constraints. But I intend to build the argument that economic conditions do finally determine the overall direction of frontier development in Brazil in the long term, and within this final theoretical framework of explanations the political activity represented here mediates those determinations on the one hand, and, on the other, the widespread violence and human suffering seen on the frontier.

5

Law and lawlessness on the frontier and the problem of bureaucratic inertia

The situation of dual authority in the frontier regions, where ownership of land may change not only by commercial contract but by the decisions of the courts leads inevitably to legal confusion at the local level. In the states of the south the Federal State leases land where the local state titles it; these leases, and even simple applications for title – which have no legal validity whatsoever – are negotiated as if they were titles (Gomes 1969). More recently in the states of the north virtually any document issued by the Federal land agency (including tax receipts and farm surveys) are sold in the emerging market for land. Moreover, the body of law governing access to and occupation of land is hugely complex, and the innumerable laws, decree-laws, regulations, instructions, injunctions and 'explanations' have never been collated. It is said that in Brazil there are in force some 120,000 laws (Stefannini 1977), and the only one missing is the obligation to obey them.

Quite apart from these legal complexities both Federal and local administrations have proved incapable of controlling the process of occupation of the land. Local land offices were not even equipped with maps of their respective regions, so that title-holders could claim land where it suited them, or extend the boundaries of their claim at will. As observed in the last chapter the land department in Pará in 1959 did not have a map of the state, let alone survey maps of its different regions, and with only three surveyors and one jeep there was no possibility of surveying the land that was titled. Fifteen years later the department was little better equipped (the Federal agencies now had survey maps but the state agency was still without) and met the flood of applications for land from large capitals in the south by dividing the state into rectangles with no thought for the topography of the areas so titled. This and similar situations on other frontiers led to titles being 'superimposed' one upon the other, and left the land departments open to corruption by local politicians and outside economic interests. And in the Federal areas of Amazônia, the legal limit on title to land of 3000 ha (which with the legal obligation to hold half the land under forest reserve leaves but 1500 ha for exploitation) was a positive incentive to politicians and capitalists alike to corrupt the bureaucracy and capture more land than their legal due.

Administrative deficiencies create legal difficulties, and administrative practices compound them. It is the practice in Brazil to register all titles in the local land offices (called *registros*). Although this is a purely formal practice, the registering of a title is often interpreted as proof of its validity. This can create complications if the local state cancels titles which continue to be registered in the land offices, or issues titles which are never so registered. In addition, the registering of true title is often confused with other formal practices, such as previous 'inventories' of claims to land – again having no legal validity – carried out during the Empire and First Republic (ITERPA 1976). These claims (which may be no more than an assertion of *posse* which was never ratified by the law) may come to be 'transferred' several times over the generations – in the local land office. For these and other administrative reasons it becomes difficult to distinguish true title to land from claims which are mere historical curiosities.

As already suggested the legal doubts presented an open invitation to economic interest groups – inside and outside the local state administration – to manipulate the law in order to get control of the land. The states themselves are not intimidated by the legal confusion from issuing titles; on the contrary, the confusion encourages what is often an indiscriminate but profitable titling policy. Where ownership of large areas is sub judice (dual authority), one title seems as good as another. At the political level, titles issued by the local states may indeed represent a challenge, a political reply to the series of prohibitory or confiscatory decrees issued over the years by the Federal State. In the poorer states, early in the history of occupation (Mato Grosso in the fifties, Pará in the sixties), the titling may also be a means of buying political support or paying political favours. But the principal motive for titling is economic gain. And while only the states' land departments can issue titles, the pressures to do so will often originate outside the states' administrations.

Economic interests which wish to speculate in land must needs establish some claim to it. Even in frontier regions where 'ownership' of land becomes a fluid concept in a context of constantly shifting legal advantage, this is best done by means of title to it. So control, if only temporary, of the region's resources will depend on the complicity – at times the connivance – of the local state administration. The economic groups must exert sufficient political pressure to have the state issue titles in their favour, or decide, in the courts, in favour of their titles rather than others. They can then use the legal sanction of the title to legitimate their control and promote their economic interest. In this way, with the Federal agencies embroiled in long lawsuits and the endemic administrative inadequacies, economic interests can operate fairly freely, and speculate in land with impunity.

The titling of land in Paraná, Mato Grosso, and Pará

In Paraná, the state issued titles to ambitious economic groups in the west, such as those of Dalcanale and Bento Gonçalvez. Typically, these groups would sell up to double the area of the titled land, in the confident hope that the state would find more land to cover the sales. But in some areas all land would already be titled (for instance, there were upwards of thirty 'colonising' companies operating in the south-west of the state in the late fifties) (Relatórios DGTC n.d.) and the only way to meet the demands of the groups is to 're-baptise' the land, and sell it again. In short, to 'find' land which did not, in fact, exist. This was the major impulse to the pernicious practice of double and treble titling, by which two or three estates might 'legally' exist under different names, but cover approximately the same area of land. Lupion, governor in the late fifties, practised blatant 'double baptism' on land claimed by both private enterprise and the Federal State (Praefeitura Municipal de Palotina 1971); and when he and Dalcanale fell from being fast friends to bitter enemies he titled on land already titled to Dalcanale. Although such 'double baptisms' took no account of previous titles, Federal leases, or even developing colonisation projects, they were a feasible short-term solution, as the new 'owners' would rarely leave the city to take possession of the land. But as land prices rose they would sell it again – to other colonising companies or to individual peasants.

In Mato Grosso the public lands of the south were liberated by the Federal cancellation of the Laranjeiras contract, and, with the 're-democratisation' of the country in 1946 and the dissolution of the Federal Territory, the rush for land began. The local state constitution was changed to allow individual applicants to buy up to 10,000 ha (the limit had previously been 500) and state politicians, economic groups from Rio Grande do Sul and São Paulo, and the Laranjeiras company itself, competed to amass large estates. The favoured mechanism was that of bogus applications (*procurações falsas*), whereby fictitious persons applied for land in order to avoid the legal limit. As the companies and groups were selling at ten times the price of the state, speculation took hold until the opposition party (the PSD at the time of the early fifties) contested the issues of titles in the

Federal Congress, and the whole region was placed sub judice. This stemmed the issue of titles temporarily, but in the late fifties – this time with the PSD in power and João Ponce at its head – speculatory titling started again. As in Paraná in the same period, demand for land from outside the state was growing, but the local state administration titled to its own politicians. Adherents of both parties appeared involved in the search for personal profits. Political careers were expected to be self-financing in Brazil, and poor state administrations such as that of Mato Grosso had to avail themselves of their major resource – the monopoly of land.

Obviously the titling by the state did little to modify the conditions of monopoly. The only small modulation possible within the confines of Brazilian bourgeois democracy (although similar effects are observable much more recently in Pará, in the period of 'military authoritarianism') were the colonisation projects that were rushed through in key areas, immediately prior to an election. After the election these projects would be allowed to languish. Interestingly, this ploy never actually contributed to winning an election: the principal parties (UDN, PSD) have always alternated in power since 1945, and, despite the imposition of new labels on old party bottles, this tradition continues until the present day.

In Pará the story was much the same. Small-holders, companies and economic groups were in the rush to buy land, and if it was already 'private' property there was no limit to the area that could be bought in any one transaction. But many of the titles were false or contested, and again the state administration might be responsible. Since 1964 it has been less easy for the local state to indulge in 'double baptism' and other similar devices – the Federal State having on at least one occasion forced the resignation of a governor (Leon Peres in Paraná) allegedly engaged in dubious land deals – but state land offices, and state judges may still promote speculation. One example was that of the local land office in Tucuruí which, without any legal authority, 'titled' the land along the state road from Tucuruí to the Transamazônica, and bribed a local public prosecutor and a local police chief to collaborate in the fraud. These titles appeared valid, but were, in local parlance, 'cold documents' (*documentação fria*), and such titles derive from the continuing corruption in the state, and also Federal State, departments.

The indiscriminate issue of titles by the states is essential to this 'bureaucratic' speculation in land, but it would be misleading to suggest that the bureaucracy is uniquely responsible for the speculation. Many private interests take advantage of the legal confusion to forge their own documents, and titles to land, in order to take control of the frontier. In other words, deliberate cheating over land is added to bureaucratic malpractice and to the anyway tangled legal histories of these regions; and certain regions, such as Paragominas in Pará, are notorious for the incidence of forged titles. The forgeries are carefully accomplished, and, in the case of 'old' titles, the paper carefully yellowed by 'age'. The speculators, or *grileiros* (practising *grilagem*) can often find willing accomplices in the private real estate offices which flourish wherever frontiers expand. These offices, manned by a peculiar breed of parasitic land lawyers, are as much interested in profit from land as the speculators themselves. They may have pretty premises, and a head office in the south, but their business is buying and selling land on the frontier. And because their lawyers are busiest where conflicts are thickest on the ground, they are as happy to sell false titles as any other. Stories in Pará tell of such lawyers who have worked within the Federal land agencies, and, having created sufficient land problems through the bureaucracy, then set up their own offices in order to solve them. *Grilagem* contributes to the legal conflicts which afflict the frontier regions, and while it is a predominantly private practice, could not continue without the at least occasional collusion of the state and Federal administrations. At the Federal level, INCRA in the three years before 1975 fought many legal battles (involving some 6,000,000 ha) to recover public land from the *grileiros* in Maranhão, Pará, Acre, and Amazonas (*Estado São Paulo* 1975). But given the size of the bribes to which INCRA personnel are exposed, no-one knows how many cases go uncontested; while the second major Federal agency in Amazônia, SUDAM, continues uninterested in the validity of titles presented with cattle and mineral projects (Schmink 1977) – such that forged and irregular titles may be underwritten by Federal fiscal and credit incentives. So *grilagem* strengthens the tendency for the same land to be sold several times over.

The double baptisms and *grilagem*, the bureaucratic and private speculation, generate the multiple and conflicting titles to land on the frontier. The title-holders are not, needless to say, farmers, but financiers, industrialists, doctors, merchants, and politicians. If they have to do with the land at all they are 'colonising companies', which wish to profit from their title by selling land to the peasants moving onto it. These title-holders may never leave the city to see the land to which they hold title, which was true of all but a few of the 550 benefiting from a recent colonisation scheme in Conceição, Pará (Ianni 1977). They are the 'asphalt farmers' (*lavradores de asfalto*) engaged in 'Coca Cola colonisation' (*colonização Coca Cola*). As this peasant name-calling suggests, titles are held by those far from the land, while the peasants cannot make themselves recognised as rightful owners of the land by their use of it and need for it.

The claims to land, which these conflicting titles contain, represent the possibility of speculative profit. The speculation spawns a multitude of lesser legal offspring, which are the hundreds of local law-suits between holders of 'old' and holders of 'new' state titles, between holders of state titles and holders of Federal leases, and between these and holders, wittingly or unwittingly, of forged documents. In these conditions all appearance of legal normalcy breaks down. Every possible piece of paper laying claim to land is registered and negotiated as title; instances are recorded of children as young as three 'applying' for title to land, and having their signature duly 'witnessed' in the land office; fictitious persons also fill in many applications; and colonising companies spring up which are nothing more than a pile of papers which are shuffled occasionally to produce quick profits. Where speciously legal sales of land to peasants do take place, the contracts can contain so many restrictive clauses that they are bound to default at some stage – and the land be taken from them. Not surprisingly the many legal disputes, and the frustration engendered by the legal delays (local cases may run on for anything from five to ten years; major cases involving state and Federal administrations may go on for decades) provoke political conflicts at all levels – between peasants and peasants, peasants and title-holders, title-holders and title-holders, or any of these and the state administration, which, as we shall see, finally impinge violently on the real world of the frontier.

Thus the legal situation becomes *doubly* confused: just as two or more estates may be 'superimposed' one on the other, so within these larger areas individual titles may be decked, one on top of the other, building what is known on the frontier as 'two-' or 'three-storey' land. The result is that title to the land is no real guarantee of 'ownership'. Moreover, title does not represent the only right to land on the frontier, which may also be legally claimed by the fact of its occupation or *posse*; and with titles to land appearing increasingly invalid, de facto occupation rather than legal documentation begins to define the process of frontier expansion. In other words, the rights of the title-holders are further contested by the activities on the ground of the *posseiro*. These activities, it will be recalled, were seen to be divided into two types in the dominant ideological view of the frontier process (Chapter 1); this discussion is now extended and recast in the complex legal context of the frontier to reveal something of the social reality underlying the ideological categories.

Posse and the posseiro

In Brazilian law, as was noted at the beginning of the last chapter, there are two distinct ways for lands which are public property (*terras devolutas*) to become private property: either by force of a legitimate title, issued by local state or Federal State, or by virtue of the effective appropriation (*apossamento*) of the land by individual initiative. Local states and Federal State issue title when land is purchased, or for the purposes of colonisation; occasionally land is also leased – by the Federal State in the frontier strip, and by certain local states, such as Pará, to control more closely extractive activities (rubber, cashew nuts). The right to land through its appropriation, called *direito de posse*, is made closely conditional on what is called *cultura efectiva e morada habitual* – a stock phrase meaning that before this right can be transferred to a property right the settler, or *posseiro*, must legitimate his claim by living on the land and cultivating it. All this was stated as early as 1850 (in Law 601 of that year), and the democratic Constitution of 1946 (article 156), and the decree-laws of the 'revolutionary' regime of 1964, all reiterate the rights of the *posseiro* (Stefannini 1977). Indeed the Land Statute of 1964 (Decree-law 4504) insisted in different articles (principally 12, 13, and 20) on the social function of the land (Messias Junquiera 1966), and, in keeping with this, Decree-law 200 of the same year emphasised the legal rights of the *posseiro* to a 'family plot', which should take priority over the sale of land, whether by state or Federal governments. Even where those occupying the land cannot clearly demonstrate their *direito de posse* (which is established only after one year and one day) they should take priority in its purchase – and if they have improved the land they should be able to buy at a lower price. In short, by law, the *posseiro*, on public lands, should be able to get full title to the land he claims by demonstrating that he works it.

In practice, things are different. The *posseiro* must 'petition' the local state administration, or Federal agency, to get title to the land, and this proves near impossible for the peasant farmer – without resources, unversed in the ways of the bureaucracy, illiterate. Laws passed in Pará in the late sixties insist that he must bring complete personal documentation, and a complete description of the land he claims – location, name, boundaries, area and topographic details (Ianni 1977). Later these requirements were expanded to include a 'rational development plan'. Moreover, prices were quoted, against the spirit of Federal legislation, that appeared cheap to economic consortia, but proved beyond the resources of the peasants. And, in addition, a tight schedule for payment, including a prior deposit in the state bank was established. In other words, the *posseiro* was clearly prevented from 'petitioning' for land by bureaucratic manipulation, and this situation is typical of the *posseiro's* position vis-à-vis the local state administration (big companies, on the contrary, have the bargaining power: the Ometto group was quoted Cr$870 per ha by ITERPA, the state land agency, but finally paid only Cr$100). In this regard, if the land in question is under Federal control, as it is likely to be in Amazônia today, the *posseiro* may find the Federal land agency less intractable; but INCRA, and before it IBRA, will issue a series of 'survey documents', 'occupation licences', 'letters of assent' etc., before considering the issue of a title to the land (Seffer 1976).

However, the legal possibilities and bureaucratic impediments facing the *posseiro's* appropriation of land, and its legalisation, are merely the niceties of his position. Overriding them is the fact that nearly all land in Brazil is titled before it is occupied, or, failing that, during the process of its occupation. The peasant who is *posseiro* moves to the frontier in the search for land, but this land has been appropriated in law before he can appropriate it by his labour. The *posseiro* on the ground cannot, of course, distinguish between what might be public and what private property, and by his migration to the frontier may therefore be transformed, unwittingly, into an 'invader' (*invasor* or *intruso*) of private property. As such he may be seen as asserting his de facto rights against the legal rights of the title-holders: indeed, as ideological categories, the terms *posseiro* and *intruso* are sometimes used interchangeably on the frontier. But it is important to keep a measure of neutrality for the term *posseiro*, meaning a peasant who attempts to establish *posse* – despite legal complications that he cannot comprehend; the frontier peasant who deliberately invades private property is then called *intruso*.

What can be seen to happen on the frontier is that the peasants adapt to the realities of their environment, and act accordingly. The awareness grows that legal appropriation has preempted their struggle for land. So wherever a legal dispute puts the ownership of an area in doubt, it is highly susceptible to 'invasion' (*intrusagem*) by peasants who are, no doubt, taking advantage of the situation. In this way the continuing legal confusion is a powerful incentive to the *posseiro* to turn *intruso*. In this sense the dual authority and multiple titling in frontier regions effectively promotes a predatory and precarious pattern of land settlement. Legal confusion causes social disruption. And it is now apparent that legal disputes over land between state and Federal administrations, and between these and private economic groups – owners and pseudo-owners – draw more fuel to the fire in the form of *intrusos*. While justice prevaricates over rights to land (as it did in Chopinzinho for example in the west of Paraná, in a three-way law-suit between Federal State, local state and the Dalcanale group) (Relatorios 1970 a and b) the commerce in pseudo-rights and rights of *posse* intensifies. If all titles to land are in dispute, it is, in some sense, still public.

Posse and the posse industry

Posse can be a profitable undertaking – a sort of poor man's speculation. The *posseiro* moves onto the land, stakes his claim, perhaps clears some land, and then sells his claim to another peasant pushing hard on his heels. He then moves on to repeat the process (or alternatively, if he genuinely wishes to work the land, he may claim a larger area and call friends and relatives to join him on the frontier). What he sells is not so much the product of his labour – though the clearing of the land may enter his calculations – as his *direito de posse*. *Posseiros* who live by staking out the land in this way arrive first on the frontier, and will be the first to move on. An IBRA survey of the Andrada estate in the west of Paraná showed 3316 families of *posseiros* within the area in July 1967, and in another survey in 1971 4500 families were counted – but less than 200 of the families on the estate in 1967 were still there four years later (IBRA 1967a; INCRA 1974). Almost all the first levy of *posseiros* had sold out and moved on.

The frontier is not, of course, a clearly drawn 'line of advance' but a process of irregular settlement within a region, which is progressively defined internally. It fills out and it fills up. The *posseiro* will gain more from his stake as land prices rise (which is also the result of his labour), so he will want to operate close to or within the settled areas of the frontier. When this commerce in *posse* turns systematic, and concentrates not only around the settled areas, but especially in disputed areas, then it is known as the *posse* industry (*indústria de posse*).

As land prices rise, the different title-holders – owners and pseudo-owners – begin to assert their rights, and legal dissension ensues. It has already been argued that this provides an incentive to the *posseiro*, but many argue too that the 'invasions' of disputed areas are directed. The direction may come from lawyers, who see profit to themselves in the protracted legal battle; from 'owners' who see the greatest gain in the possibility of expropriation in the face of social conflict; from lumber-merchants, who may send in the *intrusos* to get out the wood from land to which they do not have title. If a title-holder then files an injunction to prevent the wood from leaving the area, the lumber-merchant may organise the *intrusos* to protect his interests, which he will present as theirs too, and may even hire a lawyer to help them in 'their' fight.

The variations on this sort of theme are many, but of the different possible connections with 'invasion' that of lumbering appears the most frequently. In Pará, the many invasions in Conceição do Araguaia – (and especially in Redenção) – were associated with the upward of one hundred sawmills set up to cut the rich mahogany reserves; and when the military police moved in to quell the revolt in São Geraldo do Araguaia, in the same municipality, the first step they took was to close the lumber company. In Paraná, the 'colonising' companies, and especially CITLA, when forcing the peasants to buy the land they worked, wrote 'contracts' forbidding them to cut the native pines – so creating private reserves of pines for the company, through what anyway was an illegal claim to the land they stood on. And the very folklore of this frontier recognises the connection in such sayings as 'where there's trees there's trouble' (*onde tem pinheiros tem brigas*): in this lore *mato* and *macho* are seen to be mutually suggestive. In effect, 'invasions' are more frequent as the frontier moves through its 'pre-capitalist' stage (Chapter 2), and to the degree that they are directed will be so by entrepreneurs engaged in the extractive activity so typical of this stage. In places it is relatively easy to trace the progress of the *posse* industry, following the focus of conflict and 'invasion', whether through the south-west of Paraná, or along the Cuiabá-Santarém in Pará. But it is finally always open to question just how far the 'invasions' are criminally engineered, and how far they express the unorganised struggle of the oppressed peasantry on the frontier.

Posse and capital

It is correct to suppose that in most instances of confrontation on the frontier the small *posseiro* faces the large landowner or capitalist entrepreneur. This is what occurred on the Santa Rita de Cassia estate in the west of Paraná (Shigueru 1972), which was bought by the RIMACLA company in 1967 (Agropecuária e Indústria RIMACLA Ltda.), with the ambitious plan of transforming the forest into artificial pasture for cattle-raising. It had applied to the BNDE (Banco Nacional de Desenvolvimento Econômico) for financing, but as the loan delayed the *posses* multiplied – from just four families in 1967 to some 260 in 1970. RIMACLA had begun to clear the forest, and even its own employees were selling *posse* to the peasants. When the army refused to help the company, it employed its own armed men to 'invite' the *posseiros* to the company HQ. In

June 1972 four company men died in a shootout with the *posseiros* and those responsible fled to Paraguay.

The case is instructive because there was no legal doubt over ownership. In law, confrontation of this kind can be reduced to the basic breach between property rights or title, and property possession or *posse* (title defining what is legally right and *posse* the real situation): they 'should' coincide but with the legal confusion and rapid occupation of the land on the frontier, rarely do. The reaction of the *posseiros* here, which seems unexpectedly quick and violent, is explained by the violence they suffered two years previously. The public prosecutor in Foz do Iguaçu had called in the military police to deal with small cases of invasion. By the time they arrived the *intrusos* had gone, but local landowners seized the opportunity and paid up the police to clear their own estates, and many genuine *posseiros* – some there from seven to ten years – were scoured from the land. These *posseiros* had now settled again, and paid for the land (to Federal leaseholders and 'real estate' lawyers from Espirito Santo). Mixed with them were *intrusos* who had seen IBRA decide in favour of the *posseiros* on a neighbouring estate, and now held out for the speculatory profit which a Federal expropriation would bring. Where the Federal State expropriates, title goes to the *posseiros*, who can then sell at a profit (on occasion the owner himself will hold out for expropriation, preferring to receive compensation for himself, and leave the problems to the State).

For these reasons the *posseiros* proved intransigent when the representative of INCRA, the Federal land agency, came to speak with them. He promised plots of land in Mundo Novo (Mato Grosso) or in Amazônia. The *posseiro* version was that they had been promised 'a new world' (*um mundo novo*) and, as for Amazônia, INCRA might as well offer land there to the company, and divide up the estate for the *posseiros*. They were aware that if they showed enough fight they might create a sufficient 'social problem' to extort an expropriation from the bureaucracy. In the event, after legal wrangling, they were finally moved off the land, as is usual in such cases. But the story demonstrates the antagonistic relation of title and *posse* on the frontier: just as *posseiros* can be scoured from what is a legitimate claim, so, occasionally, they can invade land which has legitimate title. This implies that on the frontier both title-holders and *posseiros* may initially be more prone to speculate than cultivate, but, if so, the *posseiro* is gambling not for profits but for his livelihood. Finally *posse* is important as a physical occupation by those without land, of land already too much titled.

This confrontation between small peasants and large landowners or capitalists has occurred on every frontier in Brazil, and continues today on an expanded scale in Amazônia. The *posseiro* by definition has no title and traditionally the large owner does. But the real struggle is not over legal right but over land, and it is a struggle carried on between very unequal forces. As we shall see, the real dimensions of the struggle are revealed in Amazônia today, where the rush for land has been so intense that even large capitals have often not yet succeeded in getting title to land. In 1970 39% of the area of landholding in Amazônia was in effect *posse* (as against 14% in Brazil and 34% in Pará) but over half this area was occupied by 'properties' of over 10,000 ha. As many of these 'properties' enjoyed financing from the Bank of Amazônia and benefited from SUDAM's fiscal incentives, it becomes clear that the question is not finally one of 'legality' but of economic strength and political power. The tens of thousands of *posseiros* continue to fall victim to forces beyond their control. To them, the language of law – a structured context of communication for those with capital – is pure mystification.

Law and violence on the frontier

The slowness and bias of the legal system, and the very concept of 'ownership' of land, work against the peasants on the frontier. Legal right to land always belongs with others, and their own 'legal' claims and protests are always invalidated in the mysterious language of law and bureaucracy. The role of law on the frontier is perfectly clear: if the land does not belong to those who occupy it, then it can be sold to them; if they cannot buy, then they can be moved on, and the land sold to other peasants arriving at the frontier. Or, more simply, the land may be legally appropriated once it has been cleared by peasant labour. And it is peasant labour which creates the value from which speculative profits are drawn. So that while individual migrants may reap temporary benefits from the legal confusion and proliferation of *posse*, the peasantry as a class is mystified and exploited. Finally, where the provisions of the law are incomplete or insufficient, then violence is used to force the peasants from the land, or sever their surplus from them.

The many conflicts over land, bred by the situation of dual authority, multiple titling and *posse*, create a high potential for violence in the frontier regions, but this potential is only precipitated, in most cases, at the moment when the real appropriation of value begins. The title-holders, and companies, which have fought for legal control of the land, will force the peasant to pay for it, or will force him from it (so 'cleaning' the land, in Brazilian usage). To these ends they employ hired gun-men (*jagunços*), or the police in their off-duty hours (*mata-páu*) – or even on duty, and out of uniform. Whether these 'cleaning' operations are backed by a court order or not, they do not take account of the genuine *posseiros* who have occupied the land for some years, and the more resistance they meet with the more violent they are likely to be. The title-holders responsible for this callous clearing may not even know where 'their' land is situated, which serves to demonstrate how these conflicts, which are, in their majority, generated *outside* the region, only find violent expression *within* the region – where houses are burnt, animals slaughtered, women abused and men shot. Every peasant on the frontier can recount the crimes of violence he has known or suffered.

Most aspects of the violence are illustrated by the operations of the colonising company CITLA in the south-west of Paraná in 1957 (see Chapter 1). CITLA had gained illegal 'title' to lands in the south-west by chicanery and corruption at the time of intense migration into the region. Its first step was to sell some of this land to its own subsidiaries, and to companies from Rio Grande do Sul, and then it began to sell to the peasants themselves. As they did not have the capital to buy the land outright, they were forced into signing 'contracts' (which CITLA's boss Lupion could then discount against his debts in the Bank of Brazil) – 30% down and the remainder over two years. The peasants were reluctant to buy. Not only were the titles issued by CITLA worthless (*'o documento não tinha fé público'*), but in many cases they had already paid two or three times for the land they worked – to rival title-holders in the same region. But they were convinced by the company's gunmen who travelled the countryside in groups of three and four, armed with Winchesters and machetes, and threatening crops and family with fire and guns.

The contracts were imposed in the areas of native pine forests, and, as already observed, specified that the trees remained the property of the company (again, illegally). Where there were no such reserves, as in Capanema, the APUCARANA (a CITLA subsidiary) gave the peasants ninety days to pay or leave – on the clear expectation that they would go. The presence of surveyors marking out the land left no doubt as to the intentions of the company, which were to clear the land and sell it anew to moneyed migrants from the north of the state and São Paulo. Here the peasants reacted, and people died in small-scale gun battles between peasants and the surveying teams. Even in the forested areas of Pato Branco and Francisco Beltrão those peasants who could not pay were driven from the land. There was no quick way out of the immediate vicinity except by bus, and some could not afford even that. The very poorest peasants were left to bear the brunt of the companies' onslaught.

The lack of legal definition of the land legitimates and complements the violence. Moreover, the very fact that all titles are 'temporary' means that the capital gains on the land – a series of forced 'levies' in the case of CITLA – must be realised as quickly as possible. Hence CITLA's gangsters in the south-west pressed harder for immediate payment on the land once it appeared that the imminent legal judgement on true title to the land would go against it (in a strangely analogous way, the state land agency of Pará, ITERPA, is today pressing to clinch as many land sales as possible, regardless of their legality, in the expectation that the public lands remaining to it will shortly be expropriated by the Federal agency, INCRA). In fact, as soon as CITLA began to bring violent pressure to bear on the peasants, the opposition deputies of the south-west (those of the PTB) began to advocate expropriation by the Federal administration – in the public interest (*por interêsse social*). Once this won general approval, CITLA despaired of winning the land legally, and so intensified the campaign of terror in order to realise a quick 'profit'. Peasants were shot down in the forest trails, and then flung into the river, or buried. 'Sensationalist' newspaper reports told of shallow graves in wheat fields being planted over with potatoes – that the gun-men might exhume proof of their work. Toward the time of the revolt it is thought that some two or three people were being killed every day – and there were enough gun-men in the region to make this possible. In other words, while the legal ambiguities create the conditions for speculation in land, the rapid realisation of profit requires violent incentives, and a cruel coercion of labour.

Bureaucracy and the land problem

In different periods the different local state administrations have attempted to 'solve' the problems created by conflicts over land – but with little success. In Mato Grosso the land department itself was closed on three occasions in an attempt to contain the corruption. The first time was in the late fifties, when it soon opened again under pressure. The second time was in 1961 – a time, according to the man who closed it, when all titling was controlled by six or seven economic groups allied to the state administration, and the state Ministry of Agriculture little more than a real estate office. On that occasion the legislative assembly demanded it be re-opened before they would approve the state budget (and their desire to complete their own deals was all too evident). Yet the indiscriminate dealing began to cause severe social conflict in certain areas like Cáceres, and the department was closed for the third time by the governor Pedro Pedrossian in

January 1966. The move earned him untold enemies, and during the rest of his term he was several times denounced to the Federal authorities for 'subversive' activities. In the early seventies INCRA requested that the department open again – not to effect further sales, but to classify the confusion. Its archivists are now at work.

In Paraná, such were the conflicts sown by multiple titling in the west, the state administration itself had to expropriate land (that is, expropriate its own title-holders) in certain areas in order to pacify them. None of these expropriations took place while Lupion was in office, of course, but all arose from his malpractice of 'double baptism'. One such area was the Piquirí estate in Palotina (Praefeitura Municipal de Palotina 1971) (a highly fertile municipality about 60 kilometres from the Paraguayan border) which was transferred to the Pinho e Terras colonisation company. The company carried forward an exemplary colonisation scheme, only to be stopped by Lupion's double baptism and the bloody conflicts between the gunmen of his title-holders and the company's colonists. The Federal army's Fifth Frontier Regiment had to move in to pacify the area. As legal conflicts continued the state later expropriated the area in 1963, but insisted that the colonists pay for the land (again) at the price fixed in the expropriation order – effectively protecting the state's title-holders at the expense of the peasants. Moreover, the state did not take the elementary precaution of surveying the scheme before expropriation, so inviting invasion by *intrusos* looking to take advantage of the dispute. Other expropriations, like that in Chopinzinho in 1965, brought similar results.

In Pará all sales of land were suspended in April 1975 (ITERPA 1976). Here, as elsewhere, administrative inadequacies meant that land in some areas was titled several times over, but the state administration faced the additional embarrassment of ignorance regarding even the approximate location of very large land sales. The state land agency ITERPA was set up some months later to find a solution to the problem, but it was poorly organised, ponderous, and has proved incompetent to resolve the intricacies of the legal problems. Its priority was to review the titles issued by the state itself, and in particular the 185 titles which had been issued fraudulently during the years 1967 to 1972, and which had already been annulled (*Liberal* Nov. 1976). These titles, which covered some 750,000 ha, could not be investigated as the Secretariat of Agriculture had allegedly destroyed the only register of state lands in Pará in order to protect itself. So ITERPA resolved to re-sell these areas to the original buyers – but at higher prices! Its solution was therefore eminently commercial rather than legal: to sell back the land to the *grileiros* so as to make as much profit as possible in the shortest possible time.

Thus it appears, perhaps not surprisingly, that local state attempts to resolve the legal confusion met with failure. But it must be remembered that the Federal State claims many of the lands in the frontier regions – those in the frontier strip and the new security strips to either side of the roads in Amazônia – and asserts its legal right to control the occupation of these lands, in the interests of 'national security'. It hopes to achieve this by means of the Federal agencies operating in the local states. But far from fulfilling its aspirations, the decrees it issues and the legal decisions its courts hand down tend to deepen the land problems by invalidating local states' titles so creating further incentives to invasion and uncontrolled dealing in land. The land agencies for their part have proved expensive and generally ineffectual: the Land Statute of 1964 raised the hopes of many, but since early in 1965 expropriation 'in the public interest' has been delicately dropped from the agencies' priorities by continual 'postponements', and cadastral surveys and taxation have been made their principal tasks (Min. Agricultura etc. 1968).

The Land Statute made IBRA responsible for settling the land problem of the frontier strip, which – following advice from USAID personnel – was divided into different 'land districts' (*distritos de terra*) in Rio Grande do Sul, Santa Catarina, Paraná, Iguatemí (Mato Grosso), Corumbá (Mato Grosso) and Rondônia. The districts were directed and staffed by military men, with the brief to survey and title lands wherever disputes were in evidence. This policy met with partial success in some districts, but certainly not in Paraná. In the first place, despite the great variety of land problems found along the strip, and throughout Brazil, IBRA was centralised in Rio, and the extreme lack of coordination between the centre and the localities led rapidly to the discovery that 'in practice, theory is different'. In the second place, every land problem is finally a social problem – which was especially the case in Paraná where the fate of thousands of families depended on IBRA's sensitive intervention – but neither IBRA, nor its sister agency INDA, nor any other Federal agency in the region had the capacity, or the flexibility, to deal with these problems.

By the early 1970s it was estimated that about 250,000 out of a total of 300,000 titles in the west of Paraná, issued by the state and private companies, were secure (Costa de Albuquerque 1970). Yet the others remained in dispute, and open confrontations, deaths, shootings, were still likely in at least ten per cent of the region. Yet INCRA, the latest and largest of the land agencies, endowed with military organisations and command, spent its time debating obscure points of law. It remained tied by legal debates in the courts, and compromises with the state administration. By the Constitution it could in principle cancel all titles and distribute the land anew (Gastão de Alencar 1971), but with one and a half million people farming in the region this would not be a solution so much as a social catastrophe. On the other hand, an 'engineered' policy of selective expropriations ran into the legal paradox of expropriation of land to which it anyway asserts full legal rights!

The endless legal debates and the preoccupation with procedure eventually drew the agency into an apparent inertia. The legal debate over land, which continued to create antagonism between local state and Federal administrations, was now academic in its complexity, but it has already been observed (Chapter 4) how crucial the outcome of the debate was in financial terms. The local state fought every case to prove the validity of its titles – for it had to pay compensation wherever they were annulled; similarly the Federal State paid compensation if it was obliged to expropriate. During the early sixties the local state – fearful of a compensation calamity – pressed for as many individual, 'out-of-court' settlements as possible (it is claimed that 50,000 cases were solved in this way between 1961 and 1964) (DGTC 1966), but this piecemeal solution at local level was halted by the judgement of the Supreme Court in 1965 (cf. Chapter 4) and the increasing interference of the Federal agencies. Today INCRA itself is returning to this formula, which has the advantage to the litigants of finishing with the business of law and the costly legal games (*jôgos de advogados*) which may delay a decision for decades. They are attracted by a compromise which gives something now, in place of a prize that may never materialise. Interestingly, in working to this end, INCRA clearly renounces any principle of legal right for the notion of 'fair shares'. Finally the question of legality is revealed as spurious to the political and economic struggle for control of land on the frontier; the law outgrows those who hoped to profit from its manipulation, and it is rejected in favour of a percentage of the prize.

These settlements provide a practical, but only partial solution. In the majority of cases INCRA must await the outcome of the lengthy law-suits still in progress. In the meantime it continues with the purely formal exercise of ratifying the titles in the frontier strip which are not in dispute. In other words, INCRA's 'solution' to the problems of the frontier depends on the whim of individual title-holders and the slow turning of the years; it appears resigned to the 'reality' of the situation. But it is difficult for INCRA to defend its inaction as a policy decision. On the contrary, the very nature of the 'solution' clearly reveals its incapacity to formulate policy and *impose* it on the frontier. It is INCRA's promise that it will 'fix the man to the land', but this promise is the leitmotif of every land decree and land law of the past thirty years, and one that has never been fulfilled. Legislation has not been lacking, but, despite declarations of intent, no Federal agency or department has been able to implement it. Even with an amplified administrative apparatus and 'revolutionary' changes at the level of the Federal State, the bureaucracy seems incapable of resolving the legal conflicts and social problems of the frontier regions – and this is as true of Paraná in the fifties as of Pará in the seventies. There is a wide discrepancy between the stated goals of Federal policy-making and the Federal agencies on the one hand, and, on the other, what actually happens on the frontier. The frontier remains impervious to Federal policy.

But this judgement on the bureaucracy must be adjusted by a broader perspective within which this 'imperviousness' is only apparent. The frontier is finally determined at the level of accumulation within the national economy. Since the thirties this accumulation, it was noted in Chapter 3, has partly depended on a high rate of exploitation of peasant labour, and a continuing transfer of value from countryside to city. The frontier contributes to this accumulation, and the trajectory it traces will finally be interpreted as a cycle of accumulation. This cycle can appear 'independent' of Federal policy and the operation of local state and Federal agencies, just as these remain incapable of 'directing' frontier development. But, in fact, the frontier is only independent of the stated goals of policy and the institutional criteria of the agencies insofar as these deny the dynamic of the cycle of accumulation. Moreover, the legal confusion which afflicts the frontier, and the open conflicts which attend it, no less than the bureaucratic 'inertia', are crucial for the creation of the conditions of this accumulation. Later the analysis will look to the relations of accumulation on the frontier (Chapters 8 and 9); for the moment it focusses attention on these conditions of accumulation created by legal and political penetration of the frontier.

Bureaucratic inertia and the cycle of accumulation on the frontier

As argued from the first chapter, the frontier experience is the expression of a violent struggle for land, and this struggle is mediated by the operation, or, better, inoperation of the law and the bureaucracy. Legal confusion breeds political conflict and social disruption. Justice is not done and is certainly not seen to be done. In the face of this turmoil the bureaucracy recoils from radical action and retreats into inertia. In its attempts to mount administrative solutions to what are political problems, it becomes weighed down by the workings of its own apparatus (cf. Chapter 7). Characteristically of authoritarian States the bureaucracy is caught in that 'strange combination of Reichstaat and arbitrary power, of slow legalistic procedure and military command style' (Linz 1964), which deadens the impact of its policies on the frontier. But it is precisely the legal delays and bureaucratic inertia which mediate the struggle for land. The failure of the law to establish legal right and the failure of bureaucracy to implement formulated policy effectively *prevent* legal resolution and allow the struggle to continue for as long as it favours the cycle of accumulation.

The struggle is a class struggle fought over the appropriation of surplus and the appropriation of the land, and both forms of appropriation enter the cycle of accumulation on the frontier. Legal delays and bureaucratic inertia are catalysts of this accumulation by mediating the struggle so as to prevent legal resolution in the short term, and so as to promote a particular pattern of resolution, or private property, in the long term. In other words, individually, the continuing conflicts express the real competition for land between title-holders, individual and corporate, and between these and the peasants; but collectively they play an important role in the cycle of accumulation and in the slow and painful emergence of private property. While the correspondence is not complete, it is useful to recall here the distinction between the 'pre-capitalist' stage of frontier expansion and the final 'capitalist' stage. It is certain that the lack of legal resolution favours accumulation through extractive activity and speculation in land characteristic of the 'pre-capitalist' stage. But accumulation in the 'capitalist' stage requires not only private property per se, but also a particular pattern of land tenure. Bureaucracy, both administrative and legal, mediates the emergence of this pattern. As implied in Chapters 2 and 3, accumulation on the frontier finally requires a highly concentrated pattern of land tenure which, in practice, means the reproduction of the dominant duality of minifundio and latifundio.

The emergence of minifundio is illustrated clearly in the case of the south-west of Paraná, where CITLA's operations and the subsequent revolt and 'victory' of the peasants led to administrative chaos and a complete lack of control over the occupation of the land. In the wake of the revolt waves of migrants broke upon the region from the south, and *posse* proliferated. Within four short years the region underwent a metamorphosis into minifundio (GETSOP 1966). Significantly, the Federal bureaucracy only reached agreement on action *after* the transformation. GETSOP (Grupo Executivo das Terras do Sudoeste do Paraná) arrived in the region in 1962, to begin surveying and titling. In short, it legalised the regime of minifundio at the very moment when the region began to produce agricultural goods in some quantity for the national market. Private property in minifundio favours the transfer of value through market mechanisms characteristic of the 'capitalist' stage.

If anything the fragmentation of land-holding in the south-west accelerated with the arrival of GETSOP. More *posseiros* moved into the region in pursuit of a 'government' title, and the rising prices for land which came with the increased security led many peasants to sell part of their land, their *direito de posse*, to the new arrivals. GETSOP officials freely admitted that there was no question of redistributing the land more 'rationally'; the peasants who had paid several times over for their land (to *posseiros*, to colonising companies, to the state government) would accept no 'remodelling' of the land tenure situation. Even bureaucratic attempts to prevent further fragmentation met with failure. A decree forbidding the sale of land for five years after receipt of title was, predictably, ignored and as these 'illegal' land transfers created more conflicts the '5 Year' rule was dropped. The bureaucracy had to accept the fact of the fission of the land into minifundio.

But this regression to a regime of minifundio is only one side of the question. In most cases the combination of conflicting titles and appropriation through *posse* leads to a highly concentrated pattern of land-holding, which implies the presence not only of minifundio but, overwhelmingly, of the large landed estate or latifundio. This is confirmed by the changing picture of private property in Conceição do Araguaia in Pará (as shown by IBGE data quoted in Ianni 1977). Over a period of just twenty years, from 1950 to 1970, although the number of *posseiros* increased seven times over (from 292 to 2136) the percentage of the total occupied area in *posse* drops from 91 to 43. The percentage of land-holdings with title did not change over the period, remaining constant at about 5%, but by 1970 they covered 56% of the total occupied area. Although these figures define an increasing concentration they do not indicate its full extent: the average area of *posse* over the period also increased from 82 ha to 136 ha. Further figures from INCRA (1974), already quoted in Chapter 2, show that in 1972 there were 961 latifundios in the region, comprising 58.8% of the holdings and covering 97.46% of the area. However, in that same year, title-holders took but 60% of this area, so demonstrating that at least 30% of the total occupied area was in *non-titled* large landed estates. These latifundios are themselves *posses*. At the same time, there was a dramatic drop between 1970 and 1972 in the absolute number of *posses* (from 2136 to 1341), indicating that the smaller *posseiros* were being pushed off the land. In other words, the concentration of land, in the transition from the 'pre-capitalist' to the 'capitalist' stage on the frontier, was proceeding faster than its 'legalisation' in private property. Only when concentration of land is complete is a full regime of private property finally imposed.

The picture drawn from Conceição do Araguaia is illustrative of the contemporary process in the whole of Amazônia, where concentration of property in land is proceeding apace. This concentration is not only rapid and recent, but is thrown into stark relief by the traditional leasing of land to land-holders in the region through such systems as *aviamento*. Significantly, under pressure from private enterprise, Brazil's new Civil Code denies local states the legal right to lease land. It is inescapable that the concentration of ownership comes in response to the range of Federal fiscal and credit incentives designed to promote the 'development' of Amazônia. Property clusters under the control of the big capitals – the banks, the enterprises from the south, the multinational corporations. But while the speed and scope of the process of concentration in Amazônia is particularly striking, it must be understood in the more general context of the cycle of accumulation on the frontier.

On every frontier the contentious transition to a regime of private property in land accompanies the transition to the 'capitalist' stage of frontier expansion. In the 'pre-capitalist' stage accumulation has occurred in the relative absence of property relations and these are imposed when accumulation can no longer continue without them. Necessarily, and this is the point of characterising the different stages, the relations of appropriation and accumulation will then be different. In the south-west of Paraná the regime of minifundio is legalised at the moment of the region's integration into the national market; the surplus of the petty commodity producers could then be appropriated largely through the price mechanism. In the west of the same state a firm legal definition of land-holding follows investment in infrastructure and easier access to the region, where land in selected areas is 'released' to the larger producers. In the south of Pará, the rapid concentration of property holding follows the influx of capital seeking high profits in raising beef cattle for the international market. The evidence suggests that, just as legal conflicts and confusion contribute to push the frontier through its full cycle, so the legal resolution contributes to fix the regression to minifundio, or latifundio–minifundio, or capitalisation and concentration within select areas – at the moment when such resolution best accords with the requirements of capital accumulation.

Thus law and bureaucracy, by their political intervention into property relations on the frontier, successfully mediate the struggle for land. But this is not the end of the story. In the first place, although they mediate the struggle, they do not finally determine it; determination occurs through the economic relations which define the classes and class fractions in struggle. In the second place, this intervention in property relations does not describe the full range of bureaucratic mediation on the frontier. As suggested in Chapter 1, it is not, in general, entirely satisfactory to separate the 'economic' and the 'political' in this way, and the separation is no easier in the process of concentration of private property and transition to capitalism on the frontier: for instance, areas of concentration and capitalisation experience rapid accumulation which provides the incentives and resources to settle outstanding disputes within and around the areas in question; while areas with firm title and secure tenure attract capital; and large owners can buy out small *posseiros*, and small capitals be absorbed by large ones. With this in mind the argument now proceeds to pull the 'economic' and the 'political' closer together in a discussion of the relationship between bureaucracy and private enterprise – in the context of colonisation on the frontier.

6

Private and public colonisation of the frontier and the pattern of bureaucratic entrepreneurship

In explaining the process of frontier expansion, the emphasis has rightly been placed on spontaneous migration to the frontier, rather than on planned colonisation of the frontier. The pioneering movement is primarily determined, as was argued in Chapter 3, by the 'surplus' of labour and monopoly of land in the countryside, and by the rising demand for staple foods in the cities. Planned colonisation appears to have contributed little to the movement, not because plans for colonisation were lacking, but because most such plans, both public and private, met with failure. Public planning of colonisation has recurred throughout the period of the pioneer frontier, with occasional concerted efforts to direct the pioneering movement, such as the series of incentives to colonisation provided by Vargas under the Estado Novo, which were heralded collectively as 'The March to the West' (Esterci 1972). But of the tens of national colonies inaugurated in the years following the collation of this legislation in 1945, only one or two managed to survive and prosper (Diegues 1959), which is indicative of the small success of public colonisation initiatives over the period. Such public policies necessarily obeyed political imperatives, such as 'peopling the political boundaries of the nation', and therefore the collapse of most of the colonies may be explained by a lack of any economic viability. But private colonisation plans, which in principle should respond to clear economic incentives, appear to have met a similar fate. There have been hundreds of 'colonising' companies operating on the pioneer frontier over the period of its expansion, but their achievements in colonisation have been conspicuous by their absence. Yet, public attempts to colonise the frontier continued, and the colonising companies operating on the frontier increased in number. Reviewing the period as a whole, what is remarkable is not merely the failure of planned colonisation, but the continuing commitment of resources, both public and private, to colonisation.

Undisturbed right to land is probably a minimum requirement for colonisation, and it is relatively easy to explain the failure of most colonisation projects, both public and private, in terms of the legal disputes and social strife so prevalent on the frontier. The successful exceptions must have escaped the general conditions of the struggle for land presented in the previous two chapters (as will be seen below). But within this perspective any account of colonisation will simply extend the previous analysis into a discussion of a predictable record of failure. It is not that this record is not interesting in itself, but its relevance to the advance of the analysis lies in its apparent contradiction with the continuing commitment of economic resources to colonisation. Investigation of this commitment will show that while the public projects and private companies may not achieve colonisation, they contribute to accumulation on the frontier. Thus, while actual colonisation achievements do little to shape the pattern of frontier expansion, the study of colonisation programmes and companies is important in revealing the operations of private enterprise on the frontier, and their relation with the State bureaucracy. In short, 'colonisation' provides another context for investigating the forms of political mediation of the struggle for land and accumulation on the frontier.

Colonising companies and the stages of frontier expansion

The economic logic of colonising operations in general nearly always implies some form of extractive activity, and especially lumbering, in order to off-set the heavy short term costs of colonisation (Roche 1968). But it is recognised that the majority of companies confine their activities to extraction, and the rapid and speculative sale of land to other companies, or to the migrants moving onto the frontier. These companies are either not prepared to invest or not prepared to await the return on an investment, but are looking for windfall gains. This strategy distinguishes them from the minority of companies which plan their projects as long-term investments, and are prepared to wait for the profit that will accrue from agricultural production and progressive integration into the market. These two kinds of company represent different types of economic enterprise, and the predominance of one or the other on any frontier corresponds to different stages of frontier expansion.

The lumber and land speculation companies are predominant on the frontier in its 'pre-capitalist' stage. Evidently the relationship between economic enterprise and the production and property relations existing on the frontier is mutually reinforcing. On the one hand, the companies respond to the prevailing insecurities of tenure and title, which work to prevent investment and promote a highly predatory pattern of occupation; on the other, the companies seek to speculate, and favour the lack of definition of property rights which propagates *posse* and unstable settlement. While peasants are pushed off the land as fast as they occupy it, these companies will continue to reap profit through legal chicanery and violence. But with the transition to the third stage, and the emerging regime of private property, these companies tend to disappear from the frontier.

In this 'capitalist' stage, the land development companies can come into their own. The emergence of a firm land tenure structure and integrated market and credit networks creates the conditions for a high return on their initial investment. But evidently these companies cannot arrive 'after the event'. Colonisation implies a relatively equal distribution of land to the frontier colonists, and the legal confusion, speculation and violence of the preceding stages contributes to reproduce the dominant duality of minifundio and latifundio on the frontier. Therefore, if such companies are to succeed they must arrive early at the frontier in order to preempt the legal confusion and prevent occupation of the colonisation project by *posse*. As mentioned above, the general conditions of frontier expansion discourage colonisation, and only in this way can the 'exceptions' escape the conditioning.

Probably the greatest exception in the history of the pioneer frontier is the Companhia Melhoramentos Norte do Paraná in the north of that state. By buying up individual titles and acquiring land from the state before the rush for land began, the company became full owner of 1,236,000 ha as early as 1927, and succeeded in almost eliminating the usual legal disputes and social conflict (Melhoramentos 1975). The company then proceeded to survey and sell land, providing a full range of agricultural and social services to the colonists. As the colonisation continued, so the railway from São Paulo was extended, new towns springing up at each new location of the rail-head. Although this colonisation was largely completed in the forties and fifties, today the north of Paraná remains the most prosperous agricultural region of its size in Brazil.

No colonisation carried out on the frontiers under study has managed to emulate the success of Melhoramentos Norte, but the MARIPA project became legendary in the west of Paraná, as had Melhoramentos in the north. MARIPA (Industrial Madeireira e Colonizadora Rio Paraná) bought some 290,000 ha known as the Fazenda Britannia from Madeiras Alto do Paraná, an English lumber company with its headquarters in Buenos Aires (Oberg 1957). The sale took place in the forties, following Federal legislation during the Second World War prohibiting foreigners from owning land in the frontier strip. MARIPA effectively controlled this land, and surveyed it with the intention of colonising it with migrants from the south and from São Paulo. Early operations were financed by the sale of pinewood, and the migrants themselves were not slow to arrive: by 1956 6074 of the 9618 available plots had been sold (Oberg 1957). Roads were built to accompany settlement, and marketing networks established with their centre in the Empório Toledo, a merchandising company set up in 1950 under the joint control of MARIPA officials and colonists. Within a few years local produce was being exported direct to the freezer plants and forwarding markets of Ponta Grossa and Curitiba, and large wholesalers, including national companies, built their own storage facilities in the area. Within a relatively short time, the area saw the consolidation of a prosperous colonisation community.

Just as Melhoramentos Norte is an exception in Brazil, so MARIPA is an exception in the west of Paraná. But outside of the MARIPA area occupation did not proceed everywhere by means of *posse* and in conditions of legal confusion. Other companies, and especially those of Luis and Alberto Dalcanale, attempted to anticipate and guide the process of settlement. The Dalcanales arrived in 1950, following some twenty years' colonising work in Santa Catarina. Wherever they could they bought title to the land, and tendered it for colonisation to smaller companies; these would operate as part of the Dalcanale group, which retained a controlling share of the capital. Many of these titles were purchased from the state, and not all were free of legal complications, but, in partnership with Alfredo Pascual Ruaro, the brothers claimed to have colonised some 300,000 ha. There were other companies of some importance in the west (Hoffman in the south-west, Soldati to the north) but the Dalcanale colonisation clustered in the area that would benefit from direct access by asphalt to Curitiba, the state capital.

These companies arrived early and so succeeded in escaping some of the instabilities created by speculation and legal confusion. But by arriving early they entered, perforce, into direct competition with lumber companies, in an environment still empty of legal norms and institutional sanctions. Where the Dalcanale group arrived a little 'later' the state had already begun to title and so the 'legal monopoly' of the land was lost; even though MARIPA arrived when the frontier was still in its 'non-capitalist' stage, later in its history Lupion tried to 'double baptise' one corner of its territory. Any legal dispute constitutes an invitation to *intrusos* to contest the land with the colonists, and the pattern of settlement for all these colonised areas was distorted to a degree by conflicts of this kind. In the Dalcanale case plans to develop commercial and industrial enterprises along MARIPA lines had to be dropped. In the light of this 'competition' it is apparent that early arrival at the frontier was a necessary but not sufficient condition for successful colonisation. In addition, the companies required the resources to defend their rights to land.

The resources in question are political not economic: the struggle for control of land, whether provisional or permanent, is a political struggle; to win it the companies had to command political support outside the region, and, usually, inside, the local state or Federal administration. This emerges from a consideration of the competition faced by the 'exceptional' companies which carried through long-term colonisations. But beyond being a specific requirement of these exceptions, such political support is a general requirement of economic entrepreneurship on the frontier. In general terms, the two types of economic enterprise in competition on the frontier attempt to appropriate value in different ways, and, in effect, belong to different fractions of capital. On the one hand, appropriation takes place through speculation and extraction; on the other through the institutionalised markets for land and goods (the different modes, both separately and in combination, are considered in detail in Chapter 8). But both fractions compete to control the land for longer or shorter periods of time, and to achieve this political control they require political representation at the level of the local state or Federal administration. It will be seen below how 'competition' and 'representation' create conflicts within the State. For the moment, the question of political supports to private enterprise is pursued in the context of the public colonisation projects.

Local state colonisation projects

It has already been observed how the land grants to railway companies in Paraná, and the leasing of land to an extractive company in Mato Grosso, did nothing to contribute to colonisation, but, on the contrary, prejudiced the colonisation plans that were to come. Similarly, in the north of the country, Pará's practice of leasing land for extractive industry prevented a more permanent occupation: the *aviador*, for instance, was not a landowner but an entrepreneur who had no interest in seeing the land colonised (Monteiro 1963). At a later date, however, most local states set up special departments, or 'state companies', in order to carry out colonisation. In Paraná, this was the DGTC (Department of Geography, Land and Colonisation), in Mato Grosso, the DTC (Department of Land and Colonisation), and in Pará, SAGRI (the Secretariat of Agriculture). But these departments were responsible not only for colonisation, but for titling land in general, and these two goals did not always prove compatible.

It has also been observed how the titling of the DGTC in Paraná was tailored to speculation rather than colonisation. In addition to the DGTC there was a state founded company, the FPCI (Paraná Foundation for Colonisation and Immigration) which, in its private capacity, was to promote colonisation and settle land problems which the department found intractable. For the purposes of colonisation the FPCI received from the state some 450,000 ha in its own right; elsewhere it was to command the murky margin of inconsistencies between the legal and the de facto distribution of the land. With state backing and a clear mandate to colonise the FPCI did successfully plan the colonisation of one or two areas (in the municipalities of Terra Roxa and Corbélia), but its overall impact was of a different order. Instituted in 1947, by Lupion in his first governorship, the FPCI functioned to protect state and private interests in the 'public lands' of the west from the first of the powerful colonisers like the Dalcanales, and served as another instrument of state supported speculation in land. Elsewhere the FPCI simply became lost in the labyrinthine legal history of the west (in most cases due to the Federal annulment of private titles during the Second World War), and could not offer uncontested titles for colonisation – even if it wished; these titles were sold anyway, to the 'asphalt farmers' in the city.

In the north of the country colonisation has traditionally failed to overcome the continuing isolation of the frontier regions. It has been estimated that in every Latin American country with undeveloped tropical lowlands, upwards of 100 colonisation projects have been begun in the last thirty years, with overall success in single figures (Nelson 1973). Colonisation in Pará conforms to this general picture. Apart from the early colonisation of the Bragantina area (near to the state capital) and the later Japanese colony at Tomé-Açu (Moran 1975), no state sponsored colonisation has met with much success. This does not ignore the efforts of the land department to settle some of the migrants by the Belém–Brasília road, where colonisation by small peasants occurred in certain municipalities like Irituia, but this was more a response to a spontaneous movement than a state initiative. In this case, lack of success was again due to speculation, and not only economic but also political. In fact, the majority of the 'state colonisation projects' were spontaneous settlements which were used for electoral purposes; tens or hundreds of migrants being titled arbitrarily in the hope of swinging the balance in any one municipality. This political manipulation of the migrants by SAGRI might also bring its economic rewards, but no colonisation occurred as a result. Thus, when INCRA demanded an account of the thirty or so 'projects' on record in the early 1970s, only three could be verified as existing at all (ITERPA 1976).

These 'political' colonisations were seen to occur in Mato Grosso, and it is no surprise to find them in Pará. They appear to have continued up to the present. As late as 1977 a major social problem was created in the municipality of Conceição do Araguaia by the hundreds of families (estimates varied from 300 to 1000) which flocked to the cross-roads with the Xingu (where the projected road to São Felix do Xingu joins the PA 150). It was alleged that SAGRI had promoted this 'colonisation' for the state administration in order to win the municipal elections of 1976 (and had advertised as far afield as Paraná). This was the typical manoeuvre of creating a block of votes, but not in this case to defeat the opposition party (MDB) but the opposing faction of the government party (ARENA). In the event the ploy failed, and all support for the project was straightaway withdrawn. Speculation began, and the choice land near the cross-roads was soon titled, so pushing the area of 'colonisation' further and further away, until it lay some fifty kilometres from the growing township. The migrants meanwhile were being intimidated into buying land that was to have been free. This was the most recent 'state colonisation'.

Federal colonisation projects

Uncoordinated attempts at colonisation had continued from the birth of the First Republic, but the bureaucratic history of Federal colonisation begins with Vargas' 'March to the West' (Esterci 1972), and not until 1954 was the first land agency founded. INIC (National Institute for Immigration and Colonisation) absorbed the three Federal departments which had previously divided responsibility in this field: the Council of Immigration and Colonisation, the Division of Land and Colonisation of the Ministry of Agriculture, and the National Department of Immigration of the Ministry of Labour (Diegues 1959). This agency, and all subsequent ones, was subordinate to the Ministry of Agriculture, and was intended to plan and execute colonisation by peasant farmers, and to promote land grants through the local states to guarantee reserves for long-term colonisation purposes. Moreover, it was to supervise all colonisation work by private companies. But INIC's performance fell short of these plans. On the one hand, the Federal colonies continued to flounder and were often occupied by a greater number of *posseiros* than colonists (as was true of CANGO in the south-west of Paraná); on the other, only 28 of at least 150 colonising companies operating throughout the country in 1957 were registered with INIC (as the agency complained in a statement to the Federal Chamber of Deputies in that year) (*Estado Paraná* 1957) – which augured badly for State control of private colonisation.

In view of its poor performance INIC was replaced in the early sixties by SUPRA (Superintendency of Agrarian Reform) which was followed after the Revolution of 1964 by the dual land agencies, INDA (National Institute of Agrarian Development) and IBRA (Brazilian Institute of Agrarian Reform). The latter were constituted to work in complementary fashion, and in conjunction with the National Labour Department of the Ministry of Labour and Social Welfare. But this succession of agencies led to discontinuities in administration, which prejudiced the progress of existing projects, and opened the agencies to manipulation by local state bureaucracy.

In Dourados, Mato Grosso, INIC's lack of resources halted work in the colony (Porto Tavares 1972); SUPRA had hardly come when it went, and there were demarcation disputes between INDA and IBRA. The discontinuities left increasing scope for local state speculation in land. The original Federal decree had specified 300,000 ha for colonisation, but on survey the allotted area was found to exceed this by 109,000 ha. The state was not slow to claim this area, and, in the absence of Federal supervision, much more besides, such that the colony now measures some 268,000 ha. Afterwards the state titled this area, but the new owners found Federal colonists already working the land; while, in the wake of Federal delays, even parts of the uncontested Federal colony were ceded by the state for colonisation by private companies.

But the vagaries of the bureaucratic history of Federal colonisation pale into insignificance in comparison with the brief given to the latest of the agencies, INCRA, to colonise the Amazon region. INCRA was founded in 1970, absorbing the dual agencies IBRA and INDA, and the decrees issued in the following two years not only made it the largest landowner in Brazil and Latin America (Chapter 4), but allocated it a key role in the massive National Integration Project (PIN) (NAEA 1974). Its crucial contribution to the first state was to be the colonisation of the land to 10 kilometres to either side of the Transamazônica and Cuiabá–Santarém penetration roads, and this it was to do through its Integrated Colonisation Projects (PICs). These PICs were to spearhead the colonisation of INCRA's 'security strips' with peasants from the North-East, and, in Pará, for example, were established along the highways at Marabá, Monte Alegre, Itaitúba and Altamira. They must be distinguished from the very different Land Projects (PFs – *Projetos Fundiários*), which were planned to regulate the existing process of occupation in, for instance, Conceição do Araguaia, Paragominas, Marabá and Santarém. In the PFs INCRA was intended to settle disputes by survey and by law and adjust the different de facto and 'legal' appropriations of the land; in the PICs it was to impose the pattern of settlement. Overall, INCRA's plans included not merely colonisation proper, but a progressive redistribution of the land, and a massive 1,498,914 square kilometres was defined as a priority area for agrarian reform.

INCRA's colonisation did not prove as easy in practice as it did on paper. It was soon evident that its legal powers were far greater than its administrative capabilities. The apparatus was not lacking, but it appeared too rigidly centralised to delegate sufficient authority to meet difficulties as they arose on the ground, and too inflexible to coordinate its plans with other Federal State and local state agencies operating in the region (Schmink and Wood 1978). On the one hand, the highly paternalist apparatus purported to solve everything, but the small peasant could get no response from it; on the other, certain services were neglected altogether in order to avoid 'overlapping' between agencies, and possible inter-agency conflict. This is important in view of the number of agencies assisting in the colonisation projects: in Altamira alone, in addition to INCRA, were found IPEAN (agricultural research), DEMA (Ministry of Agriculture advising service), SESP (health), SUCAM (malaria control), COBAL (food company), CIBRAZEM (storage and warehousing), MEC (education), SEDUC (education, SESI (technical centre), BB (Bank of Brazil), BASA (Bank of Amazônia) and ACAR (rural extension) (Moran 1975).

The mass of agencies led to a kind of bureaucratic overkill. The banks, for example, did not deal directly with the peasants, and it was INCRA which supervised credit applications, and payment schedules, with the result that most applications were rejected, and where they succeeded credit came without advice on how best to use it. Again, ACAR and INCRA both ran extension services, kept files and held meetings – often competing for custom and offering contrary advice. And although advice might be plentiful, there was little support for the real business of production and marketing (Bunker 1978). Add to this the sluggishness of the legal apparatus, with its apparent need to issue 'occupation licenses' and 'letters of consent' before full title, and it is not surprising that this 'fixing the man to the land' proceeded more slowly than had been planned. So much so that an opposition deputy, Carlos Bezerra, accused INCRA in late 1975 of deliberately deceiving public opinion: 'They have created an organisation in order for it not to function' (*Folha São Paulo*

Nov. 1975). It was certainly not functioning to protect the peasants and guarantee their possession of the land.

If the peasants' possession of the land was as precarious as ever it was because INCRA both failed to implement its original policies and failed to foresee the impact of its initiatives. In 1970 the agency anticipated that 300,000 families would have arrived in the region by 1975, but by 1972 a mere 5756 families had been settled (some 500 in Itaitúba, 675 in Marabá, 3000 in Altamira and 475 in Guamá) (Silva 1974), and by the end of 1975 INCRA could only claim to have issued some 35,000 titles (INCRA 1975). As some of these titles were issued outside of Amazônia, and some anyway did not touch land within the PICs, it is generous to suppose that 40,000 families had been moved by this time – a far cry from the figure first mooted. At the same time, although INCRA chose to colonise far in advance of the pioneer frontier, the impact of the projects accelerated the more than proportional spontaneous migration to Amazônia. It is likely that for every family brought by INCRA some five came of their own accord. As it is equally probable that another 50,000 families came through the many private colonising schemes, often speculatory, it is credible that a million and a half migrants overall entered the region during these years – a number that could not be absorbed economically, and which clearly overflowed the confines of the PICs. In short, following the incentive to migration created by INCRA, its projects are swamped by the influx of peasants and their 'spontaneous' colonisation.

The operations of private enterprise were not haltered nor the entrepreneurs inconvenienced by this *caboclo* colonisation. Indeed, it will be argued in the following chapter that the 'impact project', by design or default, effectively camouflaged the political mechanism for supplying cheap labour to the enterprises entering the region on the wave of fiscal incentives. Moreover, the failure to control colonisation (and, again as argued in the following chapter, policy reorientation within the State) led to a 'rationalisation' of colonisation, whereby not small peasants but large enterprise would carry it forward. Without wishing to anticipate later discussion, the change became evident from 1973, and in 1974 INCRA invited the participation of cooperatives of large owners from the south (such as the *Cooperativa de Trigo de Ijuí* in Rio Grande do Sul) in setting up Land Administration Companies, covering areas of 200,000 ha, and 420,000 ha in Altamira (*Folha São Paulo* 1974). The new plan had as its objective the occupation of 11,800,000 ha by 4000 small farmers on 100 ha plots, 1200 medium farmers on 3000 ha plots, and 120 big enterprises on estates ranging from 66,000 to 72,000 ha (*Estado*

São Paulo 1975). Almost simultaneously INCRA announced the POLAMAZONIA (Programme of Agricultural, Cattle and Mineral Poles of Amazônia) (SUDAM 1976b) which was designed to concentrate investment in certain growth poles, and justify the appropriation of large stretches of fertile agricultural land by the large companies (such as COPERSUCAR on the PA 150 highway in Pará). In other words, colonisation was now to cater directly to accumulation, and the 'people with no land' were to be left without.

The Land Projects

The Land Projects were designed to resolve the land problems which already existed – without thought for those yet to be created by INCRA's own projects. Where claims to land conflicted, claimants could go to court, but INCRA encouraged negotiated terms between titleholders and *posseiros*, in order to define the pattern of tenure with the minimum of delay and cost (Seffer 1976). But these laudable aims once again ran into a series of bureaucratic barriers. Even to confirm *posse* which was not contested the land had first to be surveyed to establish its boundaries and any 'improvements'; a certificate of official inspection had then to be issued, but only when proof was produced that the *posse* was registered in INCRA's cadastral survey, and that the ITR (property tax) was paid up to date: finally, a 'cultivation document' was required, establishing how much land the *posseiro* was able to farm, before the application could go to the legal department, which would then begin to issue its series of 'occupation licences' and 'letters of consent' (Schmink and Wood 1978). The *posseiro* had to have such a letter before he could even properly apply for title, and even after 'full title' had been issued he would ultimately need a 'letter of judgement' to confirm it. Not surprisingly only a small percentage of *posseiros* got title.

As observed in the last chapter, the State is legally obliged to recognise all *posses* of less than 100 ha if they are on 'public land', have been there for at least a year, and demonstrate 'regular occupation and effective farming'. Admittedly, all these conditions are hedged about with partially conflicting laws and constitutional articles and interpretations, but nevertheless it remains clear that even on private land the legal priority is for the *posseiro* to receive his 'family plot' (Stefannini 1977). Yet, evidently, this obligation of the State is transformed by INCRA into a series of obligations for the *posseiro*, which make it almost impossible for the small peasant to 'legalise' *posse* within the Land Projects. In contrast, there are distinct legal requirements for the 'regularisation' of *posse* which patently refer to a different order of land-holding. Like 'legalisation' this 'regularisation' requires 'regular occupation and effective farming', but this farming, while it cannot include lumbering or any extractive activity, may include cattle raising, even on natural pasture. It is true that the land must have been occupied for ten years in this case, but, equally, the limit of the claim may here extend to 3000 ha. As documents which 'prove' length of tenure are not difficult to procure for those with some capital and know-how, it is generally recognised that the interpretation of the CSN (Exposition of Motives no. 6. of 1976) which laid down these ground-rules facilitated the false claims and land-grabbing of the *grileiros*, while in no way helping the *posseiro* (Pinto 1976a): in implementing these rules INCRA supported accumulation through speculation.

Regarding the sale of land within areas of INCRA control, there were legal limits, which existed before the institution of INCRA. The overall limit on any one sale was 3000 ha, reduced to 2000 ha in the frontier strip, and only the Federal Senate could authorise exceptions to this rule. In addition, INCRA was obliged to sell the land within its domain by public auction, and (by Law 200 of 1967) there had to be public bidding before any State agency or department could buy this land (Palma Arruda n.d.). In practice, with the huge increase in demand for land, with the entry of entrepreneurs from the south into Amazônia, subterfuge and subornation were used to subvert the law, and 'construct' large landed estates. Everywhere speculators sold and *grileiros* grabbed the land that was speciously under INCRA's control. In Humaitá on the Transamazônica, 80% of the properties were in latifundio (*Estado São Paulo* May 1974), and in 1974 it was mooted that the highway might be re-routed, such was the social disorder sown by dealing in land in this strategic area. Finally, as will be explained in the following chapter, the legal limits on land sales were all but abolished.

All these observations have clear implications for the role of INCRA in mediating the struggle for land. Significantly, Elias Seffer, head of INCRA (Amazônia) in 1977, now head of SUDAM, and an adherent of the 'conspiracy theory' of politics, blamed 90% of all conflicts over land in Pará on the 'invasion' of private property (Seffer 1976). Although this theory takes no account of the many peasants who legitimately occupy land which is illegitimately converted to 'private property', it helpfully removes all responsibility for the problem from INCRA. Cases of invasion are cases for the police and the law-courts. This effectively denies any social role to INCRA, and leaves it with a purely legalistic orientation. This orientation has grown as commitment to colonisation has died. INCRA will now simply administer, literally, the law of the land. Moreover, invasion is distinguished from *grilagem*; it is the poor who create the trouble. And the record shows that recently INCRA has only adopted any social role in cases of 'social tension': in São Geraldo do Araguaia 815 'occupation licenses' were issued (Ianni 1977); it was likely that a similar solution would be found for Floresta (Chapter 1), where writs had been served to expel the 2000 families of *posseiros*. In short, its social role is reserved for those occasions when a small multitude is prepared to fight for its right to the land.

Paragominas (Pará): a case study in colonisation

Before referring this discussion of capital and colonisation to the general relationship between private enterprise and the State, it will be well to pause a moment, and once again place the elements of the analysis in a specific context. This will serve to 're-integrate' these elements momentarily and return the argument to the social history of the pioneer frontier. It will also serve as a reminder that the complex political relationships which underpin accumulation on the frontier are mediating a continuing struggle for land. Colonisation is one context where the results of this struggle are revealed – by the very record of failure.

The record of colonisation in Paragominas was no exception to this rule. Pioneer peasants had begun to arrive in the region ahead of the Belém–Brasília highway in the late 1950s, and they were followed by the first of the colonising companies: Colonizadora Belém–Brasília, Colonizadora Marajoara and the Cidade Marajoara (Keller n.d.). All three failed: the first perished in isolation; of the 135 plots of the second only five were ever titled, and the third was shortly abandoned in 1961. The Federal State announced a national colony for the region, but this was never established, and local state plans for two colonies also came to nought. The reasons for this failure to begin or complete colonisation were complex, but repose on the progressive appropriation of land by large landowners.

Even before the arrival of the peasants, in the middle fifties, entrepreneurs from Goiás, with authorisation from the state governor of Pará, had penetrated the dense forest along the river Capim, to survey and title land to buyers from Uberaba and Itumbiara in Minas Gerais. In this way Paragominas was founded, it is said, with land from PARÁ, daring from GOiás, and capital from MINAS Gerais. Later the approach of the Belém–Brasília road provoked the usual rush for land among landowners from Minas Gerais, Goiás, and Espirito Santo, and land speculation companies from São Paulo (the

Zancauer, Define, Pazzarezi and Moura Andrade) (Keller n.d.). At the same time peasants were pushing onto the 'public lands' of the region, but only to face the traditionally unequal competition of the *grileiros*, who produced false titles, and backed them with the use of force. Paragominas was *a terra do pistoleiro* where sharp practice and sharp shooting served to expand the large estates and squeeze the peasants into areas of minifundio.

With such rapid concentration of land-holding and in such a climate of violence, attempts at colonisation had no chance of success. The first prefect of Paragominas, named by the Federal government as Amilcar Batista Tocantins, himself planned to place an agricultural colony on land claimed by one of the large 'owners', Alfonso Leão. Leão did not have title to the greater part of this land. The colonisation proceeded according to plan until late in 1968 when Tocantins was shot and paralysed while attending an agricultural show. At a later rehearsal of the crime on the estate of Leão, he too was shot, and killed, by a local policeman – who swore it was an accident. These events were then caught up in the complications of the political campaign which was being waged for the first election to prefect. And no more was heard of the colonisation.

Once the shooting was over and the land divided, the pattern of land-holding could be made legal; and, in effect, Paragominas is now the administrative centre for one of INCRA's largest Land Projects, which covers an area of some five million hectares, and all or part of eight municipalities. In Paragominas itself there were some 7000 land-holdings by 1968, but only 63 with title – indicating the rapid rate of de facto appropriation of the land (INCRA Office). This situation was intensified and extended over the subsequent years and INCRA's principal task is to issue title. In 1975 the INCRA Office issued some 1800 titles, and 2100 in 1976; while in 1977 some 4000 cases were being adjudicated (INCRA Office). The local head of INCRA estimated in that year that it might take forty-years at that rate of progress to eliminate all legal problems in the region. Nevertheless, for the most part, the conflicts over land have subsided. They continued in Iritúia, where 1500 *posseiros* farmed 68,000 ha, and in Capitão Poço, where 800 *posseiros* farmed 35,000 ha; but Paragominas itself was calm. *A terra do pistoleiro* had become *o capital do boi gordo*; the frontier had almost completed its cycle.

Bureaucratic entrepreneurship: the local state connections

In this examination of colonisation as carried out by colonising companies, on the one hand, and by departments and agencies of local states and Federal State, on the other, what emerges as most important is not the operation of either private enterprise or public administration in themselves, but rather the political connections between them. Private entrepreneurs are economically motivated to claim the land, but to gain legal control of it they need political power in the form of 'representation' within the public administration; local state departments and Federal agencies are political institutions with institutional criteria which guide the colonisation initiatives, but their performance is always compromised by the pressure of private companies and capitals. Whether the colonisation is viewed from the public or private perspective, it is evident that the mediation of the bureaucracy is essential to achieving profits on the frontier.

This conclusion is consistent with those of the previous two chapters, and together they demonstrate a close identity, at the level of the local state, between economic enterprise and political power: in Lupion's Paraná between the land department and the land speculators; in João Ponce's Mato Grosso between local politicians of all parties and the colonisers from without the state; in Pará in the late sixties between the land department (later ITERPA) and the southern entrepreneurs. Yet this correspondence of interests is not always clear cut: different fractions of capital with different interests may compete for access to the administration (as the frontier moves through its distinct stages) as in Paraná; local politicians may differ in their response to the economic initiatives, as between politicians of the north and south of Mato Grosso; the public administration may simply deny support to the companies and capitals, as on one occasion in Pará in the early sixties. In short, the bureaucracy does not merely mediate an even set of economic determinations; not only are economic interests in conflict, but the bureaucracy itself is relatively autonomous in its actions and responses.

Nevertheless, the correspondence of interests at the level of the local state is sufficiently complete, at least on occasion, for this state to exert pressure on the Federal agencies in the interests of private capital on the frontier. This obviously has to do not only with the complicity of public and private interests in the local state, but with the situation of dual authority on the frontier, and Federal claims and colonisation plans, which create the conditions for potential conflicts, and demand Federal State intervention. The local state will act precisely to avoid conflict with the Federal agencies, and will attempt to prevent policy initiatives and intervention which might interfere with the pattern of 'bureaucratic entrepreneurship' on the frontier. It is unlikely to do this through formal channels of communication and representation, but will act informally with the Federal bureaucracy itself. Insofar as it is successful this activity contributes to an explanation of the paradox of Federal bureaucratic 'inertia', the inefficacy of the land agencies, and the apparent 'independence' of the frontier cycle.

These comments, which are illustrated below, refer to the immediate political context of the performance of the Federal agencies, and concentrate on the particular political connections between private enterprise and the State. They abstract for the moment from the broader issue of the political mediation of economic accumulation, as advanced at the end of the last chapter, in order to raise the question of the 'representation' of private capital within the State – in other words, the question of the bureaucratic mechanisms of the mediation. In the following chapter, the argument begins to embrace both levels of analysis in the context of the relationship between State and large enterprise in the contemporary period.

Bureaucratic entrepreneurship: the Federal agencies

Just as local state administrations are not always homogeneous organisations which act consistently and coherently, so the Federal agencies within the local state do not compose a monolithic structure which is capable of defending a uniform set of directives. Hence the agencies do not respond uniformly to the political pressures and economic interests of the local state: some agencies, or factions within any one agency, may prove more compliant at any one moment than others, and it may even happen (as will be seen in the following chapter) that the agencies enter into conflict between themselves at local or regional level. The mechanisms for applying pressure are complex, but can be summarised quite simply: the agencies are either acted upon by the local state administration (or local political groups) or themselves initiate, or at least 'invite', action through their own organisation and special interests. Both aspects of their performance are apparent in the following account of the Federal agencies in Paraná.

SEIPU, which is the agency responsible for all properties 'nationalised' by the Federal State, precedes the series of land agencies in Paraná, but can be seen as setting the pattern of their response to local economic interests and political pressures. In 1950 SEIPU registered a 'donation' of 450,000 ha (the estates of Missões and Chopinzinho) to CITLA (the colonising company belonging to the economic group of Lupion), following a fraudulent claim for compensation for a paltry debt outstanding to CITLA (*Tribuna da Imprensa* 1957). The real value of this 'donation' was at least 60 times that of the outstanding debt, and, to push the transaction through, the superintendent of the agency had to be bribed into compliance. By the constitution a special law would in principle have had to be passed to authorise the transfer of such a large estate in the frontier strip, but by further bribery Lupion escaped the obligatory entry in the Federal records (Tribunal de Contas) (Mader 1957). This deal not only differs from others in the size of the fraud and the effrontery of the execution, but finally can be seen to lead to the revolt in the south-west of Paraná in 1957, which was serious enough to provoke a national political scandal. But even following the revolt, the local state's and Lupion's interests in Chopinzinho were consistently defended, first by SEIPU, then by INIC, and finally by IBRA – and even in the early seventies INCRA's attorney-general continued to protect the same title holders on the estate.

In 1956 the Pinho e Terras company, owned by the Dalcanales and Ruaro, began to colonise estates in Chopinzinho to which they had full title (transferred from BRAVIACO, cf. Chapter 4). Despite this, or because of it, Lupion baptised these same estates anew, in the name of his son-in-law and other political cronies (and these titles were illicitly recorded in the local land offices) (*Relatório* 1970a). All this is moderately 'normal'. But when SEIPU sold the lands under its administration to INIC in 1958, these same estates were included, and SEIPU's superintendent denounced the Pinho e Terras colonisation to the Exchequer. A committee of inquiry was set up in 1961 which reached the right conclusion for Lupion's economic group: the BRAVIACO title, and therefore the Dalcanales' claim, must be annulled (Procuradoria Geral 1968).

At this juncture IBRA succeeded to INIC, but no legal steps were taken against Lupion's title-holders. In fact, IBRA now became more regal than the king, and contrived to bend SEIPU to its designs – SEIPU now proving less tractable under its new superintendent. SEIPU was suffering from lack of funds, as many industries were being withdrawn from its administration, and badly wanted IBRA to pay the sum outstanding from the land sale of 1958 (NCr$ 595,000 out of NCr$ 825,000), of which SEIPU was in fact to receive NCr$ 400,000 (Tourinho 1968). Both the President of the Republic and the Directorate of IBRA approved the payment unconditionally. But at the moment of signing for the money the new superintendent found he must accept certain specific conditions, namely, to work to annul the BRAVIACO titles in the Chopinzinho estate (and thereby affirm the validity of Lupion's titles) and to annul the sale by SEIPU in 1950 of two estates of 25,000 ha and 3500 ha to the Pinho e Terras company (the only land in Missões and

Chopinzinho which Lupion had failed to acquire by his original fraud).

The bare outlines of the story are sufficient to demonstrate the lack of coordination and cooperation among the Federal agencies: not only did Lupion's economic group succeed in getting SEIPU's support for its fraudulent titles in the first place, but one agency was then used to exert pressure on another in pursuit of speculatory profits – first SEIPU worked to persuade INIC to his ends, then IBRA to persuade SEIPU. But it should not therefore be concluded either that the Federal bureaucracy is always the passive partner in this process of cooption and corruption or that the response of the agencies is 'predetermined' in any way. The internal structure of each agency is not static, and at every hierarchical level interests of different weights are shifting position, until some temporary equilibrium is reached when a certain composition of interests imposes itself on the whole. By definition this *jogo surdo* is not dramatic in its operation or spectacular in its results, and can only be illustrated where there occurs a direct and unresolved clash of interests – which happened within IBRA in the autumn of 1969.

In 1965 shortly after its foundation, the most important department in IBRA (*Cadastramento*) was headed by César Catanhede, who succeeded Assis Ribeiro as President in 1967 (following Decree-Law 200 by which the Federal President ordered a 'revolutionary' administrative overhaul). IBRA as an agency (semi-autonomous autarquia) was subordinate to the Ministry of Agriculture, but its new President Catanhede was reluctant to recognise the authority over IBRA of the Minister Ivo Arzua – who happened to be *paranaense*, or a native of the state of Paraná. Arzua for his part turned his attention to IBRA's lesser departments, and especially the DN (*Departamento de Organização de Núcleos*), which was headed by Heleio Buck Silva, another *paranaense*, and right-hand man of Arzua. Silva's presence led to internal dissension in IBRA, and, although Catanhede appeared ready to contest the Ministry's authority, the Ministry controlled the budget. What few funds were made available Catanhede fed into survey work (*Cadastramento*) and one by one the work of the other departments, including the agrarian reform department (*Departamento de Recursos Fundiários*), was halted.

At this time the Ministry of Planning set up an interministerial group through IPEA (*Instituto de Pesquisa em Economia Aplicada*) to accelerate the programme of agrarian reform (Min. Agricultura, etc. 1968). One of the most active participants in the study group was one Dryden de Castro Arezzo, and the Minister sought, and got his appointment as head of the IBRA reform department. In the first months of 1969 Dryden began work with José Burigo, the director of the land department (DGTC) in Paraná, on the land problems in the west of the state. An agreement was signed with the state, despite all opposition, which for the first time made a solution to the region's problems seem possible. It was equally possible, of course, that this initiative might have posed a very real threat to vested interests in the west of Paraná, and not least to the specific interests of politicians in the Ministry of Agriculture, and within IBRA itself. What is sure is that just at the moment when IBRA might have contributed decisively to a solution to the land problems on this frontier, it was 'intervened' in, the reform department dismantled, and Dryden demoted.

This intervention, and others that can be documented, demonstrate the same pattern – the rapid reaction of local economic groups, allied to the state administration, which perceive their interests to be threatened. The susceptibility of IBRA, and other agencies, to this sort of pressure is epitomised in the figure of the director of its Technical Committee at the time, Edgard Teixeira Leite. Within IBRA he was second-in-command to the President, and represented the agency in the interministerial group set up by the Ministry of Planning. At the same time he was a director of the CNA (National Confederation of Agriculture) – the powerful representative body of the large landowners (Silva 1971). Many other key administrative posts in IBRA were also held by directors and employees of the CNA, and quite apart from the possibilities of internal 'sabotage', these posts provided channels for the systematic siege of the Ministry of Agriculture by the CNA and other landowning associations, determined to protect their interests from the danger of 'reform'.

Bureaucratic entrepreneurship: the political continuities

The evidence suggests that the pattern of informal power relations between local states and the Federal State remained remarkably stable over the period. The local states were run by different parties and politicians, and with the Revolution of 1964 the Federal State moved from the populist mobilisation of the Goulart government to the military authoritarianism of latter years, but the way in which interests are represented and pressures exerted within the bureaucracy seems to remain the same. In other words, the formal political organisations of State appear less important than its administrative structure. This is not to indulge in a new 'statolatry' in which a monolithic State administration proves impervious to even major changes in the political context. On the contrary, the State is seen as various sectors and diverse apparatuses, each one of which can be captured at different times by distinct economic and political groups, and this dynamic perception takes precedence over the formal characteristics of different regimes.

Looking back to the struggle for land in the west of Paraná, for instance, it is apparent that Lupion's control of the local state administration was crucial to the early successes of his economic group. The major rival, the Dalcanale group, was at first obliged to make use of conventional legal weapons, but Lupion, it was seen, could use his position to manipulate the Federal administration and subvert the legal conventions. But the pressure articulated through the local administration was applied less to the actual policy-making process than to the instruments of Federal policy and control of land; the fact that this pressure was exerted through the administration is more important than the 'open system' (limited pluralist democracy) which gave him access to power. This is borne out by the subsequent history of the struggle. After 1964 Dalcanale's star was rising fast, while Lupion's was on the wane. This was a result of Federal Supreme Court legal judgements which revalidated the titles to which Dalcanale laid claim and, as crucially, of other sources of support within the administration, namely friends in the Ministry of Agriculture and the highest military circles. Some of the top revolutionary generals spoke in Dalcanale's favour. Thus the mode of 'bureaucratic rule' does not change: whoever has access and sufficient political support – and in practice these capabilities will be confined to large economic groups – may gain positive results. However, the sources of political support may finally be highly particularistic.

It is true that the greater freedom of action which Lupion enjoyed during the first years of his second period of office (1956 to 1960) was owing in no small part to Kubitshek's need of the support of the Paraná PSD. But even the independence and influence of the local state PSD machines in general, and the PSD faction in Paraná at this time in particular, could not be translated into direct command at the Federal level. The influence was rather a 'negative capability' which might guarantee the state administration a relative immunity from restrictive Federal interference. This was of course important following the revolt in the south-west of the state in 1957. Within the state there was a strong opposition movement to impeach Lupion, and the issue was the subject of heated debate in the Federal Congress. It appears that he was 'saved' by his position at the head of the PSD, which answered Kubitshek's promise to Congress to investigate the case with a threat to secede from the national party. Perhaps as important was the unstinting support of Adhemar de Barros, leader of the powerful PSP in the neighbouring state of São Paulo. Even here, and it is a limiting case, the fundamental factor is the support Lupion enjoyed at Federal level, and not the fact that he gained access to the administration through a system of party politics. In other words, the formally representative political order existing before 1964 may simply have 'reinforced' certain aspects of the informal lines of command achieved through the administrative structure of the State.

In this connection it is not finally the 'federation' of the system which matters, nor the channelling of interest and achievement of power through the local state to the Federal State – although this illustrates the intricacies of the process (the mechanisms of mediation). 'Bureaucratic entrepreneurship' is not necessarily articulated through the local states, and economic interests may apply pressure and instigate action directly within the Federal agencies (which is evident both in the intervention of IBRA in 1969, and in the ongoing operations of INCRA in Amazônia). Hence, although it is correct to emphasise the continuities over the period, it must also be noted that after 1964 the political autonomy of the local states diminished, while the power of the sectoral, regional and functional organs and agencies, linked directly to the Federal State, greatly increased. At the same time, while in this 'closed system' the economic motivation behind informal bureaucratic representation and manipulation may be less evident, the characteristic 'bureaucratic entrepreneurship' is greatly accentuated and amplified precisely because the bureaucracy is now the only channel for cooption and corruption. Increasingly, economic enterprise and capital come to deal directly with the Federal State (although this should still never be understood as a monolithic apparatus but rather as an asymmetrical combination of apparatuses), and the State, for its part, begins to forge closer links with large enterprise, both national and international. These developments are debated in the following chapter.

Thus, in summary, economic enterprise on the frontier could not operate successfully without political supports – preferably at the level of the Federal State. These strategic supports were not necessarily achieved through formal channels of representation, but were more usually sought within the administrative apparatus of the State. It was in this way that the cycle of frontier expansion could continue 'independent' of Federal policies and of major political changes at Federal level in 1964. For these policies were formal statements of intent and the changes formal alterations in political authority. But, as observed repeatedly both here and in the last chapter, the frontier did not respond to the formal and normative criteria of the legal and administrative apparatuses. What mattered was the informal and direct representation of economic interests, both individual enterprises and fractions of capital, within the bureaucracy itself, which continued throughout the period of the pioneer frontier, and which led to the characteristic form of political mediation, called here 'bureaucratic entrepreneurship'.

7

The contemporary alliance of state and capital on the frontier and the contradictions within the State

In discussing Federal policy for the frontiers, it has emerged that a glaring discrepancy exists between the formal objectives of the State, and the real development which takes place on the frontier. In Chapter 3, it was observed how Federal minimum price and credit policies, which in principle might prevent the transfer of value from the frontier, can, in fact, promote it; in Chapter 5, it was seen how Federal promises to solve the land problems of the frontier were never fulfilled, or not in time to benefit the peasants on the land. This gap between policy making and policy implementation is clearly revealed on the frontiers, which offer a different perspective on the national political process from that available 'at the centre'. This perspective may correct possible misconceptions of the centre view: at the centre it may certainly be seen what is said, but on the frontier it is seen what is done. The discussion has referred to the difference between the two in terms of the apparent 'independence' of the frontier cycle from policy making in the Federal state.

In explaining the gap, there has been reference to the tendency within the authoritarian State to treat political problems as if they were simply administrative, and to the vulnerability of the State administration to the political pressures of economic groups, operating inside and outside the local state land departments. Here it has been necessary to distinguish between the legal structure and formal roles of the constituted authorities, and the informal patterns of power which are imposed by the manipulation of the bureaucracy. It became clear at the end of the last chapter that such a distinction leads to an emphasis on the continuities in the political process before and after the Revolution of 1964. The differences between the elective federalism of Brazil's 'experiment in democracy' and the military authoritarianism are not of course dismissed, but similar difficulties are perceived, if not in the passing of laws, then in implementing them. In short the continuities are carried within the bureaucracy itself. But it was also indicated that a significant shift takes place in the relationships of the bureaucracy with the economic groups operating on the frontier: while the informal patterns of power achieved through the Federal administration are not displaced they are no longer articulated through political parties and local state apparatuses, but applied directly within the Federal State. Private capital is now seeking direct political purchase within the Federal State, and the State administration, on its side, advances and multiplies in order to meet and promote the increased 'bureaucratic' activity of private capital. Consequently, this consideration of the role and scope of the Federal agencies on the frontier will concentrate on the post-1964 period and, in particular, the operation of the agencies in the Amazon region – where the State has instigated and invested in the appropriation of the land by private capital.

The Federal agencies

The presence of the bureaucracy on the frontier takes the form of many different Federal agencies, or *autarquias*. These are semi-autonomous bodies, first created by Vargas in the thirties to care for key sectors of the economy, and which since that time have continued to grow in number, size and importance, almost without interruption. Oil, steel, coffee, pinewood, sugar and alcohol have all been controlled by the agencies, and certain of them, like PETROBRAS for example, are recognised as more politically powerful than many ministries. Of course it is the succession of land agencies (the first again being created by Vargas in 1954) which is of particular concern for the study of the frontier, but in the contemporary context of Amazônia, INCRA, for all its power and possessions, must vie for influence with a continually increasing number of agencies.

The growth of the agencies has contributed to create a consensus around the principal themes of Brazilian historiography since the 1930s, those of centralisation, bureaucratisation and militarisation. These themes represent political tendencies which have certainly continued throughout the period, but which have greatly accelerated since 1964 (and here, as in other respects, the difference between the two periods appears to be one of degree rather than kind). In other words, while the process of State formation, and, in particular, the penetration of the economy by the State, is not immanent in any one regime, it comes near to being so in the bureaucracy. Politicians, who are public and partisan, may rise and fall, purge and be purged; the ranks of the bureaucracy not only remain intact, but continue to expand. But at the same time there is a prima facie plausibility in the argument supporting the inherent compatibilities of military regimes and bureaucratic cadres – hierarchy, specialisation, professionalisation – and it is true that the bureaucracy has never expanded faster than since 1964. This was an almost automatic consequence of a military regime which looked to legitimise its rule through planning growth, regulating the economy, and expanding the public sector. As in other States in different periods (Egypt, Greece, Pakistan) there emerged a familiar pattern of military–bureaucratic alliance (Feit 1973), which pursues an avowedly 'technical' and 'apolitical' programme of government. In these circumstances it is perhaps predictable that the principal characteristics of bureaucracy (as containing the informal patterns of power within the State) not only will not change but will be accentuated.

The Federal administration expands through the creation of more and more regional and sectoral agencies, which progressively reduce the sphere of authority of the local states. Traditionally the Federal State could not intervene directly in either local states or municipalities, but only indirectly through its 'coordinative' and 'supplementary' powers; since 1964 this State has restricted the power of local states through the extension of its own 'intermediate' powers, and principally through the creation or transformation of the regional agencies (Cardoso 1972a; Lourdello de Mello 1971). Illustrative of the process is the suppression of SPVEA (Superintendency of the Plan of Economic Valorisation of Amazônia) and its substitution by SUDAM (Superintendency of Amazônia) in 1966 (Cardoso and Muller 1977). SPVEA had been founded in 1953 by Vargas, but despite a constitutionally guaranteed percentage of the Federal budget, and a brief to coordinate the work of all agencies in the region, it had produced little in the way of results other than a series of somewhat utopian five-year plans, which were dutifully presented to Congress. SUDAM on the other hand had the power to coordinate Federal action directly through agreement and contract with public and private agencies, and through its own financial agent, BASA, Bank of Amazônia, which was created in the same year (Mahar 1979). It disposed of two per cent of the Federal budget, three per cent of local state budgets, a variety of interests and credits, and income from its own property; it benefited from a range of tax exemptions, and could contract loans at home and abroad. It also enjoyed access to the funds of FIDAM (the Foundation for Private Investment in the Development of Amazônia) which itself received one per cent of the annual Federal budget. But more important than its recurrent funding were the specific projects for specific sectors, all of them assessed by 'technical criteria', which it promoted through its policy of attracting private capital, both national and foreign, by the offer of tax and credit incentives (SUDAM 1976a) (and the deposits deriving from the fiscal incentive schemes swelled its own resources). The incentives allowed fifty per cent of any investment to be discounted against tax, rising to seventy-five per cent for investment in BASA bonds, and one hundred per cent if the project was established before 1971 (later extended to 1974). The incentives will operate until 1982, and are available for any project in industry, commerce, agriculture and cattle raising, falling within the SUDAM region and approved by SUDAM itself. In short, SUDAM achieved its impact through privileging private capital in the development of Amazônia (BASA 1967).

Like all regional agencies SUDAM is subordinate to MINTER (the Ministry of the Interior) and must obey its criteria for regional development (MINTER has a set of guidelines called 'the working plan for the elaboration of regional development plans'). For the business of establishing priorities for private enterprise, and assessing the projects, SUDAM's deliberative council includes representatives from MINTER, the military, the Bank of Brazil and the National Economic Development Bank (BNDE). At the same time SUDAM must cooperate in public planning, and especially in State funded investment in infrastructure. This it does by participation in PIN (the Plan for National Integration 1970) which is specifically designed to finance infrastructure investment in the SUDAM and SUDENE (Superintendency for the Development of the North-East) regions (NAEA 1974). This was the fund (some two billion cruzeiros from the Federal budget, fiscal incentives and national and foreign loans over the first three years) which was to finance the Trans-amazônica, and Cuiabá–Santarém highways (and a massive irrigation plan in the North-East) – and also colonisation and expropriation of land along these highways. PIN could also count on the support of a series of lesser agencies founded shortly before its own inauguration (Muller and Brandão Lopez 1975): RADAM (Radar Project of Amazônia 1970), the mapping agency for the rational exploitation of regional resources, and incorporated into PIN, although subordinate to the Ministry of Mines and Energy; CPRM (the Research Company for Mineral Resources 1969) which was to research into and stimulate the discovery of mineral resources, and lend technical support to the same Ministry; the CCEEA (Coordinating Committee for Energy Studies in Amazônia 1968) which was financed by the Ministry of Planning, reported to this Ministry and the Ministry of Mines and Energy, and prepared projects for execution by ELETROBRAS (the Brazilian electricity agency); SUFRAMA (the Superintendency of the Free Zone of Manaus 1967) whose aim was to integrate Amazônia into the rest of the country by the creation of an agricultural and industrial centre in its heartland – again subordinate to MINTER. In addition to these agencies the State's presence in Amazônia is extended through the Amazônian Task Force, an agency of the Ministry of Labour, whose job it is to control (some would say inhibit) the organisation of unions, and promote State unions in place of spontaneous organisations; and finally of course, INCRA, whose institution in 1970 coincided with that of PIN, and whose original objectives of opening up and colonising the region with labour from the North-East corresponded with the PIN plan for integrating Amazônia into the nation (NAEA 1975).

In no other region are there as many agencies as in Amazônia, but increased central control whether by agencies or through Federal development projects is in evidence everywhere in the interior, and everywhere influences the expansion of the frontiers. Even Mato Grosso which was for decades an administrative desert, now has its own development agency, SUDECO (Superintendency of the Centre West), modelled on SUDENE, which channels private resources to agriculture, industry and especially telecommunications enterprises. In addition, PRODOESTE (the PIN of the Centre-West) directs infrastructure investments in Mato Grosso, Goiás and the Federal district (although the resources it receives are a measure of its placing in Federal priorities: six hundred million cruzeiros in 1972 for instance, as against the two thousand five hundred million destined for the North-East) (*Estado São Paulo* March 1972), and CONDEP (the National Council for Cattle Raising Development) attempts to rationalise cattle rearing and orientates the investment both of private capital and of the International Development Bank. Then, besides the local state land department and extension agencies, INCRA intervenes to settle conflicts over land in 'priority areas' – largely areas of the frontier strip (INCRA 1970).

The great advantage that these agencies and projects enjoy over those existing prior to 1964, whether in the state of Mato Grosso, or in the state of Pará, is more resources. The resources are available because it is the Federal State which administers and collects the taxes, and its tax capability has increased dramatically over the last decade. The main sources of revenue in the countryside are the ITR (the Rural Land Tax) and the ICM (the Commodity Circulation Tax), both of which benefit local municipalities and the local state. Remarkably in Mato Grosso the budget of the state increased twenty times over from ten to two hundred million cruzeiros within six years – with the crucial difference that this budget was largely controlled by the Federal State. The State can also control and channel the resources available through its agencies, and even determine within a narrow margin the location of private investment, through its use of fiscal and credit incentives. In short the potential for promotion and control of economic activity and aid of the Federal State assures it political pre-eminence over the local states. At the same time, the State restricts the residual power of the local states and reserves itself the right, for instance, to appoint state governors. Such is the scope of civil and political reorganisation implied by these changes, and by the operation of the agencies, that it is no exaggeration to speak of a new form of political regime in Brazil (Cardoso 1975) where the relationships between State and 'society', and especially economic enterprise, are now mediated almost entirely by private and public bureaucratic organisations.

The agencies and private capital

From 1964 the State created a range of fiscal incentives to promote regional and sectoral development by private capital. The incentives were channelled through the different Federal agencies controlling the sectors and regions. On the one hand were the IBDF (forestry development), SUDEPE (fishing), EMBRATUR (tourism), EMBRAER (aircraft industry), and on the other, SUDENE, SUDAM, and a special agency for the economic recuperation of Espirito Santo (Muller and Brandão Lopez 1975). Over the same period there is a growing preoccupation with the occupation of the Amazon region (Mendes 1974). Castello Branco's 'Operation Amazônia' sought the political means to occupy this vast land area, which materialised in the form of SUDAM and FIDAM, which was to finance SUDAM's priority projects in the region. During the subsequent governments of Costa e Silva and Medici, the volume and penetration of fiscal incentives increased under the aegis of Costa Cavalcanti, the Minister of the Interior in both governments, and he promoted the concentration of incentives in the Amazon region. In April 1969 a large seminar on the Amazon was organised for the Federation of Industries of São Paulo, and in the same months Costa Cavalcanti, in a speech in the north of Mato Grosso, spelled out the State's intention of supporting private enterprise on the Amazônian frontiers (Cardoso and Muller 1977). Just as in the past the control of land in Amazônia was not by traditional land-owners but by 'renters' (*aviadores*) subordinate to the capital of the regional commercial centres and foreign industry, so in the present period an 'entrepreneurial' occupation of the land has continued – with the difference that capital no longer controls the land through the *aviamento* system, but by direct appropriation and property relations.

One year after the institution of SUDAM, further legislation orientated the agency towards cattle raising as the most appropriate vehicle for the occupation of the region. Local state legislation, as in Pará, tended to reinforce the incentives to cattle by offering a partial exemption from the ICM for cattle and agricultural enterprise. By 1968 forty per cent of the 140 projects approved by SUDAM were already in cattle and this was before Costa Cavalcanti had advertised the potential of the north of Mato Grosso and south of Pará. Over the period 1965–73 a clear preference emerges for cattle projects, which collected up to fifty-five per cent of the incentives 'released' in any one year (and never less than fifty per cent over the years 1968–73), and these were concentrated in Mato Grosso and Pará, which received 368 of the 498 projects approved by the end of the period (Ianni 1977). The pattern of concentration was actively pursued with the POLAMAZONIA plan (Programme for Agricultural, Cattle and Mineral Poles in Amazônia), whose object was to mobilise additional resources for integrated projects in certain key regions (SUDAM 1976b). Of the ten poles, the four that were destined largely for cattle projects were Barra do Garças and Barra do Bugre in Mato Grosso, and Paragominas and Conceição do Araguaia in Pará, with the specific objective of increasing the Amazônian herd to five million head by 1980 (Pinto 1976b). The POLAMAZONIA plan was a central plank of the PDAII (the Second Programme of Amazônian Development) which in its turn was but one component of the PNDII (the Second National Development Plan) (Cardoso and Muller 1977). Over the period of the plan, 1975–79, some thirty-two million cruzeiros were budgeted for direct investment in the Amazon region, while in the same span fiscal incentives were expected to generate five billion cruzeiros of investment. POLAMAZONIA was designed to concentrate these resources by region and by sector into the most dynamic activities – mining, cattle, wood. FINAM (the Fund for Investment in Amazônia) was created to direct the fiscal incentives released through SUDAM to priority sectors like cattle. Finally, to guarantee the flow of private capital, SUDAM instructed BASA and the Bank of Brazil to offer loans for approved projects at low or negative interest rates: infrastructure investment could be financed at 7 per cent over twelve years for instance, and loans for vaccine and fertiliser incurred no interest in the first five years (Muller and Brandão Lopez 1975). If the low prices for land are allowed to enter the calculations, then the available incentives might well amount to three times the sum of capital first invested. No surprise, therefore, that the flow of capital into cattle projects became a flood.

Administrative control over investment is far from complete.

SUDAM's calculations are not founded on detailed information and there is no possibility of verifying the ratio of private capital invested to that derived from SUDAM's funds, or of checking the 'costs' which enter private capital calculations. Some projects exist only on paper, and the vast expanse of Amazônia makes their discovery far from inevitable. Over the first years of the incentive scheme the different agencies approved too many projects in the competition to secure resources for their particular region or sector, and therefore private enterprises themselves had to compete for the resources necessary to their projects. As the demands of the projects far exceeded the available resources, and as it was the investors who chose where they invested, large and illegal commissions were paid by entrepreneurs eager to capture investment for their project, and brokers set up in business to manage these commissions (*Mundo* Sept. 1975). This 'open market' effectively reduced the new capital finally available for any one project, and substituted private commissions for the technical criteria which should govern project selection. At the beginning of Geisel's government (December 1974) these abuses and frauds were stopped by the simple expedient of reducing the area of choice of the investor to the region or sector of his proposed investment, and delegating the selection of the projects themselves to three sub-agencies: FISET (Fund of Sectoral Investment), FINOR (Fund of Investment for the North East) and FINAM. Yet certainly during the initial laissez-faire period it is debatable how far the fiscal incentives strengthened private enterprise in Amazoñia as against how far they fuelled financial accumulation in the south of the country.

By reducing the cost of investment capital to the entrepreneur by as much as one half or more, fiscal incentives raised the rates of profit available in the frontier regions. In other words the State attracts private capital to these regions by a direct transfer of capital from the public sector to private enterprise. At the same time the incentives to private capital, especially those for cattle projects, inspire a rapid occupation and appropriation of the land by the large enterprises from the south. Thus the Association of Cattle Entrepreneurs of Amazônia actually has its headquarters in São Paulo (Ianni 1977). The onward march of the 'cattle fever' which infected the enterprises can be traced in Conceição do Araguaia in the south of Pará. The fourteen cattle projects approved within the municipality by 1969 had risen to thirty-three by 1975 (of the eighty projects approved overall by SUDAM). The neighbouring municipality of Santana came second with twenty in that year, and by July 1976 the two together had seventy projects (Ianni 1977). The thirty-three enterprises in Conceição in 1975 put up 24.4 per cent of the total capital investment in the projects, while 71.8 per cent was supplied from fiscal incentives (and the total cost of labour, including management, in the implantation of the projects only accounted for 2.2 per cent of the total capital investment). These figures clearly demonstrate the new alliance of State and private enterprise, State and big cattle enterprise, in the private accumulation of capital. This State-promoted search for profit transforms public lands into private property (and suddenly pressures are brought to bear on the plots of the *posseiros* and the reserves of the indians; in the words of the entrepreneur, there is 'much land for few indians'). This same search results in economic irrationalities (despite SUDAM's stated objective of a rational occupation of the land) such as the 35 billion dollars worth of forest (on RADAM's calculation) burnt off the cattle land in the north of Mato Grosso and south of Pará – when there are some 950,000 square kilometres of natural pasture ready for grazing within the Amazon region (*Liberal* Feb. 1977).

The State and foreign capital

SUDAM has differed from SUDENE in that its fiscal incentive scheme covered capital from the local branches of multi-national corporations, which have been active in appropriating land in the Amazon – both with and without SUDAM support. As early as 1962 Daniel Ludwig acquired 1,500,000 hectares from the State of Pará for his JARI enterprise, and since then he has been followed by Georgia Pacific (500,000 ha), Bruynzeel (500,000 ha), Robin MacGlolan (400,000 ha), Toyomeha (300,000 ha), Volkswagen (140,000 ha), King Ranch (100,000 ha) et al. (Muller and Brandão Lopez 1975). Brazilian firms too (COPERSUCAR, 350,000 ha; CODEARA, 600,000 ha) have been successful in amassing large estates. In addition, many multinationals have opened up cattle ranches as a sort of subsidiary investment in the north of Mato Grosso, Goiás and the south of Pará (Anderson Clayton, Goodyear, Nestlé, Mitsubishi, Borden, Swift Armour) and the big national companies have again followed suit (Camargo Corrêa, Bradesco, Mappin, ELETROBRAS). This large-scale appropriation of land by big capital is measured in the average size of the cattle ranch in these regions which is some 18,750 ha. And SUDAM has begun to discriminate against projects of less than 25,000 ha.

Spearheading the multinationals' 'assault' on Amazônia (Bourne 1978) are the mining companies which have added a new dimension to the expansion of the frontier (Davis 1977). Among these companies US Steel, the company which advised on the construction of Volta Redonda in the Second World War, has played a pioneering role. In 1967 US Steel announced the discovery of a 160,000 ha iron ore deposit in the Serra dos Carajás in Pará (where deposits today are estimated at 8000 million tons). There is evidence to suggest that the discovery had in fact been made in the early 1960s, when multinationals had already begun surveying using the services of US Air Force planes operating out of Santarém in Pará. But the company had awaited the right moment, though even at the time of the announcement the upper limit on mineral concessions was 5000 ha. In 1969 US Steel entered a joint venture with the Companhia Vale do Rio Doce, the State-owned steel company, in order to exploit Carajás, and in 1969 new laws were decreed raising the upper limit to 50,000 ha with another 30,000 ha allowed to subsidiaries. These laws, followed in 1970 by news of the Transamazônica penetration road, galvanised mineral exploration, and the annual number of mineral claims in the region jumped from 2000 in 1968 to 20,000 in 1975. It is interesting to note that RADAM, which mapped all mineral deposits throughout the entire region over these years, employed the services of the Aero Service Division of Litton Industries, which belongs to the Goodyear Corporation; and that the CPRM (Company for Research into Mineral Deposits), in order to train in the techniques it needed for its 21 mineral exploration projects, had signed contracts with USAID. The big corporations continued to enjoy direct access to vital information, and also benefited from Brazilian tax incentives and State exploration assistance. By February 1972 there were over 50 international mining corporations involved in mining projects in the region and they included Bethlehem Steel, United States Steel, Aluminium Company of Canada, Kaiser Aluminium Co., Reynolds Metals, Rio Tinto Zinc, Union Carbide, International Nickel Co., and W. R. Grace and Co. In addition to its agencies the presence of the Brazilian State was maintained in the joint ventures of the Companhia Vale do Rio Doce, with US Steel in Carajás, and in the multiple Mineração Rio do Norte (including CVRD, 41%, Alcan, 19%, RTZ, 10% etc.) which was to mine the bauxite deposits at Trombetas, Pará. The State is the only possible Brazilian rival to multinational corporations which, since early 1974, have been investing over one billion dollars per month in new mining ventures (iron, copper, tin, lead, zinc, aluminium, gold, silver, chromium, manganese).

Contradictions within the State

The increase in the number and scope of the Federal agencies, and in their capacity directly to promote private capital accumulation, contributes significantly to the greater centralisation of political power in the Brazilian State. In particular, commands and incentives can now pass from the centre to the frontier, without suffering 'refraction' at the level of the local states. But the bureaucracy has had to respond ad hoc to the emergence of pressing political and economic problems, and especially since 1964 has grown rapidly and horizontally, by a process of accretion, in an attempt to solve these problems and promote growth. Hence, despite every effort to achieve greater centralisation, institutional distortions and structural deformities have not proved easy to eliminate: the State administration is far from being monolithic in structure, and, besides the occasional lack of coordination between centre and locality within any one agency, exhibits a potential for conflict between different agencies and ministries. In short, greater centralisation does not necessarily mean greater administrative uniformity, but on the contrary the increase in the number of agencies may create areas of contradictory authority, where the agencies enter into conflict.

These conflicts may be 'politically' determined at the level of the bureaucracy itself, by the institutional inflexibilities inherent in a military authoritarian regime which insists on formal roles and legal distinctions within its administrative apparatuses. This was the case with the dual land agencies, INDA and IBRA, in the frontier strip in Paraná. In broad outline their respective roles seemed clear. IBRA was to promote and coordinate all efforts towards an agrarian reform, and carry out a cadastral survey of the land. A special provision of its charter allowed it to enter into agreements with other agencies, specifically to avoid duplication of agency functions. IBRA was also to supervise colonisation projects within 'priority areas', while in non-priority areas INDA would take over. INDA, created to complement IBRA, was basically a colonisation agency and did not enter the priority areas under IBRA control. However, the Land Statute of 1964 (Article 59) clearly stated that INDA was to colonise in the frontier strip, which would reduce IBRA's role there to a cadastral one, unless the strip should be decreed a priority area. But elsewhere (especially Article 11) IBRA was made responsible for claiming back Federal lands which were illegally occupied, or unoccupied, and of course the frontier strip was Federal property (CODEPAR 1965). Thus in addition to the conflict between local states and Federal State, there existed good legal grounds for conflicts between IBRA and INDA in the frontier strip.

The conflict finally occurred in the Andrada estate (123,579 ha) just north of the Iguaçu river in the west of Paraná, which had passed by succession from INIC to INDA in 1965. At this time the land was already being fiercely contested between local state title holders and *posseiros*, who had begun to move into the area during the revolt in the south-west of the state in 1957. The rate of occupation increased in the following years and by 1967 there were well over 3000 *posses* in the area. INDA was criticised for not controlling the occupation, but, on the contrary, exacerbating the situation of illegal *posse* by building access roads into the area. But INDA, as the complementary agency to IBRA, had been forced into an increasingly marginal role of 'rural development' and lacked the power and the political resources to stop the occupation. So IBRA entered Andrada at the beginning of 1967 with a show of 'guaranteeing national security'. In the event it did nothing for over a year except conduct a cadastral survey of the area. IBRA's inaction seemed designed to delay any semblance of rational settlement in the area (which may in fact have been the case given the powerful lumber and speculative interests operating against such a settlement) and its guards were so far corrupted as to allow increasing invasion of the land, and violence to the *posseiros*. Finally INDA, its offers of cooperation consistently rejected, resolved to fight to save the area, and the majority of the *posseiros* came to its aid (*Relatório* 1968). An armed uprising followed, which, though it spilt little blood, paralysed the administration for months, and the area fell into total social disorder.

These contradictions within the bureaucracy may also be 'economically' determined by the more direct contact between State and capital on the frontier. The State seeks to cooperate with capital, but the impact of this new cooperation on specific agencies can lead to divisions between them, and the re-allocation of authority within the State. Such a division emerged in Amazônia between INCRA and its 'social colonisation', and SUDAM with its occupation by large economic enterprise. INCRA had originally been committed to the objectives of the Land Statute of 1964, which were to give the land to those who worked it. By its road building and colonisation programmes of the early 1970s it sought to settle small peasants on the land – 100,000 families immediately, and 1,000,000 by 1980 (as its head Moura Cavalcanti recalled early in 1973) (*Estado São Paulo* Feb. 1973). However, in 1973 the Ministry of Planning opted decisively for the big enterprise as the proper vehicle for the occupation of Amazônia. The Minister of Planning himself (João Paulo dos Reis Veloso) organised a mass visit of southern businessmen to the north of Mato Grosso in September 1973, in order to advertise the embryonic idea of POLAMAZONIA, and from this time SUDAM too began to emphasise the natural advantages of the region for cattle raising (Schmink 1977). In the view of the Ministry of Planning the inefficient INCRA colonisation plan had to be cut, and its effects in sectors of the government and public opinion counteracted. Moura Cavalcanti, President of INCRA, resisted the growing pressures to open areas of social colonisation to large enterprise (the constitution itself would have to be altered in order to sell areas larger than 3000 ha he argued), but his replacement in the Geisel government, Lourenço Tavares da Silva, accepted the new strategy. In May of 1974 the new Superintendent of SUDAM insisted on the incentives scheme, with cattle raising as the key sector, so that the disordered occupation along the Belém–Brasília road might never be repeated (*Estado São Paulo* 1974). One week later the new President of INCRA agreed to participate in the new process of occupation, and the Senate would authorise the necessary land sales: INCRA was not against the large enterprise, but against the latifundio, and, as proof of his intentions, the President of INCRA announced that 10,000 peasants already settled by INCRA along the Transamazônica would be 'recycled' (i.e. moved off the land) to fit them to the new entrepreneurial plan.

When beginning the work of colonisation in 1971 INCRA had consistently reserved all property rights to itself: private capital might participate in the colonisation programme, and INCRA would even guarantee a return on investment, but the land would remain public, prior to sale to the colonists themselves (Stefannini 1977). But the colonisation and road building programme itself contributed to the increase in migration (and the consequent creation of a labour force) and to the increase in land prices, both of which inspired greater interest by private capital in the region. And, once the Ministries of Planning and the Interior (SUDAM) had declared in favour of the large enterprise, political pressures built up for the conversion of INCRA's public domain into private property. The battles within the bureaucracy continued for some time, and as late as 1974 Moura Cavalcanti warned against the dangers of reproducing the agrarian structure of the North-East in Amazônia, but his warnings were ignored. In fact a little more than a year later (May 1975) INCRA announced its intention of selling large estates in Amazônia, and opened public bidding for the first 1,400,000 ha (*Estado São Paulo* 1975). It planned to ask Senate permission to sell 709,000 ha in Roraíma (487,000 of these to large companies) and 1,200,000 ha in Amazonas. By the end of that year it wanted to sell some 8,400,000 ha of the 311,000,000 ha under its jurisdiction. By October over 1000 plots of 1000 to 3000 ha covering some 2,400,000 ha had already been sold, and areas in Roraíma, Rondônia, Pará and Amazonas were reserved for future auctions (*Estado São Paulo* 1975). 2,300,000 ha of fertile red earth in Altamira, Pará, which had been divided up for INCRA colonists, were now to be sold to large enterprises. Finally, in April of 1976, INCRA obtained authorisation to sell large estates, which far exceeded the 3000 ha limit: new upper limits were established of 500,000 ha, 72,000 ha and 66,000 ha for colonisation, forestry, and cattle respectively. Following SUDAM's directives, INCRA now contemplated the sale of 66,000 of these new estates, or some 70,000,000 ha by the end of Geisel's period in office.

The decision to sell marked a watershed in the occupation of Amazônia, which was reflected in INCRA's publication *Achievements and Goals* of May 1975, where 'economic objectives' took plain precedence over 'social concerns'. The only colonisation projects for that year, three of them, envisaged settlement for some 6000 colonists on about 600,000 ha (Schmink 1977). Clearly the struggle for 'social colonisation' is lost and from now on the large estates of private capital will drive out the small and medium owners.

By the time of the second National Development Plan (1975–79) the alliance between the State and large enterprise in Amazônia was secure. At the ideological level, the key phrase which emerged was the 'social function of land', which could be translated as meaning that the land must be productive, in the sense that it must yield high returns for capital. The 'social' nature of this function could be assessed in the labour market: the 10.5 billion cruzeiros, 7.5 billion of them invested in cattle projects, estimated to be captured through the fiscal incentives scheme by 1979 were to create some 60,000 jobs, at a cost of 1.8 million cruzeiros per job. The social colonisation plan, now abandoned, was to have created 125,000 jobs in its first phase, at a cost of 32,000 cruzeiros per job (Muller et al. 1975). In short, the land goes to capital not people, and this capital may be national or foreign (today foreigners can even buy land in the frontier strip – a significant change which reveals the real content of the State's preoccupation with 'national security'). Large enterprises or latifundio – it makes little difference to the dispossessed peasant. It means monopoly over means of production and markets, so reproducing the concentration of economic and political power in the hands of the few. Very probably this concentration in Amazônia today surpasses that achieved in the colonial occupation of Brazil.

As a postscript it may be noted that while it was the Ministry of Planning which advocated most loudly occupation by large enterprise, the bureaucratic battle was fought between INCRA of the Ministry of Agriculture, and SUDAM of the Ministry of the Interior. This Ministerial conflict continued in the debates between the IBDF (Brazilian Forestry Institute), again of the Ministry of Agriculture, and SUDAM. In principle the IBDF insisted on the key Article of the Forestry Code, which forbade the removal of more than 50% of the forest cover on any one property in Amazônia – and also attempted to cajole the wood industry into obeying criteria for reafforestation. This was its brief. SUDAM, on the other hand, had ambitious plans for zoning the whole Amazonian area to plan for ordered exploitation of its forest resources, and to control this through the creation of State and mixed enterprises which would supply the wood to industry. The IBDF opposed the plan as an open invitation to foreign capital, which already owned forests, to devastate whole areas in order to reap rapid profits; while SUDAM replies that its intention was precisely to strengthen national capital enterprise in the sector. A new Forestry Code was being elaborated in 1977, which was to decide the result – but the indications were already clear: when the IBDF in Belém, Pará, prosecuted Volkswagen of Brazil for illegally devastating a huge area of forest under its control, the IBDF administration was simply dismantled. SUDAM is the agency which leads in the alliance between big monopoly capital and the State.

Interpretations of the role of the State

The intervention and operation of the State on the frontier reveals that its formal organisation is fragmented into opposing agencies and departments of State. The bureaucracy is divided into diverse apparatuses, and it is possible to interpret these as the political apparatuses of different capitals or class fractions: agencies and sectors of the State might propose goals and implement policy measures in the service of different dominant class fractions and groups. In this perspective it would be possible for instance to see in SUDAM an agency which has been captured by the interests of the multinational corporations (and which has been used to bring opposing agencies, such as INCRA, into the service of these same interests).

While such an interpretation will certainly explain something of the reality, it is perhaps too mechanistic to explain the political process which underpins frontier expansion. It has already been observed (Chapter 5) that no State agency has successfully resolved the legal and social problems endemic in frontier expansion, because to do so would be to deny the central dynamic of the national economy which derives, partially, from the kind of accumulation which continues on the frontier. In other words, legal and social conflicts on the frontier 'express' determining economic forces, and no one agency or department can be expected to reverse the direction of development. But just as the legal confusion is seen to promote a certain pattern of accumulation, so the divisions and contradictions within the State administration contribute to this same accumulation. At one level, such administrative contradictions may simply compound the very legal confusion which has tended to bind and divide bureaucracy in the first place, so creating further confusion and areas of 'multiple authority', which favour the appropriation of land by the politically powerful. The advantage to the state of Paraná in creating social tension within its own borders (Chapter 4) clearly falls into this category; this result of the legal conflicts ran directly contrary to bureaucratic objectives of 'national security' and 'internal order' as defined by INCRA. At another level, the administrative contradictions differ from the legal confusion in their contribution, by creating new conditions for accumulation. In Amazônia, although INCRA's 'social colonisation' was defeated by SUDAM's 'economic occupation', the initial intervention of INCRA was of crucial importance in providing access roads and the mobile labour force for the later entry of private capital. Moreover, in its 'social role' INCRA became guardian of 311,000,000 ha, which could then be sold to large capitalist enterprises when INCRA was won to the dominant role of the State as entrepreneur.

The State administration may be fragmented but it has a coordinating axis, which is precisely the alliance of the State and large monopoly capital (Cardoso 1975). The divisions and contradictions within this organisation are possible as long as they promote accumulation, and do not put at risk its character of dependent capitalist State. In its fundamental function of capitalist reproduction the State is a monolith. In the last chapter the political power and access to bureaucracy necessary to the operation of capital on the frontier were explored under the heading of 'bureaucratic entrepreneurship'. Now it must be emphasised that since 1964 the State itself has acquired a hugely increased capacity as entrepreneur and as a generator of enterprises – and today in Amazônia intervenes directly to promote an accelerated process of accumulation on the frontier. Since 1964 the State has not only been an organisation (bureaucracy) but an enterprise (although certain of the largest State enterprises such as PETROBRAS, ELETROBRAS and CVRD were founded in an earlier period) and a partner of large capital in an economy where the primus inter pares is multinational monopoly capital. The bureaucracy fulfils the important political function of combining and integrating these different roles, both of which are in evidence on the frontier. The decision of the Minister of Planning in September 1973 to press for the creation of big properties, under the control of large monopoly capital, in the 'priority areas' along the principal penetration roads of Amazônia, was taken in order to 'supplement' insufficient saving within the economy (the rate was 15%) by large-scale economic activity on the 'new frontiers'. The State as entrepreneur invested one billion dollars in infrastructure, so arriving first on the frontier, in order, in partnership with large capital enterprise, to raise and maintain the rate of accumulation within the national economy.

Conclusion

It is the dominant relationships of the economic system which determine the overall direction of frontier development, but this development is mediated at the political level. The institutions and structures which used to characterise this level (parties, suffrage, representation, Federal organisation etc.) have lost relevance since 1964, and in the recent period, more than ever, the mediation is achieved by the bureaucracy – in both its legal and administrative organisations. This chapter, and the previous three, have explored the means and motives of this mediation. A simple paradigm of the changes occurring since 1964 might propose that the interests of the 'civil society' are now 'represented' directly within the State apparatuses, but, as the analysis has attempted to suggest, this is a change in degree rather than one in kind. Bureaucracy and law have been the main mediators of the economic forces driving frontier expansion since that expansion began. However, the political purchase that economic interests have within the State has certainly been more secure since 1964, and this has been examined in the impact of capitals and class fractions on the reallocation of authority within the State, and in the dominant role of the State as entrepreneur and ally of large enterprise. But the mediation is not mechanical or unilateral. In the absence of an open 'political process' the State itself needs to assimilate autonomous economic groups operating within the national society, and especially on its frontiers. This it does not merely by bureaucratic 'corruption', but by the bureaucracy's role in lending support to economic speculation and in promoting economic entrepreneurship, which serves to secure a high rate of return on capital invested on the frontier. The results of this State activity may occasionally run contrary to other State objectives: in particular, military aspirations of integrating the national territory, guaranteeing national security, may be damaged and even reversed. But this is seen as a small price to pay for a continuing high rate of accumulation on the frontier, which, by funnelling surplus to the industrial and financial centres of the country, feeds the process of accumulation in the national economy. The dynamics of accumulation on the frontier are explored in the following chapter.

PART 3

Accumulation and authoritarianism

8

Primitive accumulation and violence on the frontier

The process of frontier expansion has been observed to be very violent. Indeed this violence defines the process in some degree. But the explanation for the violence is less evident. It may be seen, in descriptive terms, as equivalent to 'lawlessness' and therefore endemic to frontier regions existing beyond the reach of the law: it is random in nature, particular in motive, and criminal in its conception. It may be seen, in more moral terms, as the result of a ruthless search for gain by evil and unprincipled entrepreneurs and politicians, who do not include the costs of violence in their calculations. Finally, it may be seen in political terms, as a result of institutional incursions on the frontier: the legacy of legal confusion and conflict, and of bureaucratic bungling or inertia. Possibly all of these 'explanations' contain elements of the truth, but they all participate in partial perspectives of the total process. The intention now is to integrate these different perspectives into a more complete theoretical framework, which will contain not only these descriptive, moral and political elements, but also the economic relationships which underpin violent behaviour on the frontier.

Violence and the stages of frontier expansion

Explanations of the violence as 'inevitable' in the 'lawless' frontier regions appear to refer to the precarious occupation of the land in the initial stages of frontier expansion. This precariousness is seen to derive from the predominant activity on the frontier which is correctly described as the predatory exploitation of the natural environment in different types of extractive activity. Neither lumbering nor the collection of *caúcho* (Chapter 2), for example, favour a firm pattern of land tenure, and no permanent regime of private property can emerge during these 'non-' and 'pre-capitalist' stages. So, it is suggested, violence and intimidation emerge, in its stead, as mechanisms for resolving disputes over claims to land. But while such observations are perfectly valid as far as they go, the arguments they support are circular (of the sort, the frontier is violent, because there is no respect for private property, because there is no rule of law, because it is violent; or, more simply, the frontier is violent because it is a frontier), and do little to explain the genesis of the violence. To begin to do this, the violence must be viewed not merely in the context of the descriptive characteristics of the extractive economy, but as a direct result of the labour relations existing in the 'non-' and 'pre-capitalist' stages. In short, it is important to recognise at the outset that violence characterises not only the struggle for land, but also the control of labour on the frontier.

Violence is used as a mechanism of labour control where other forms of social control are absent. This was evident in the discussions of the operations of Maté Laranjeiras in the south of Mato Grosso, of rubber extraction in Conceição do Araguaia, of lumbering in the south west of Paraná – and nearly every serious study of the frontier concurs in this conclusion. In his study of the north of Paraná, Monteiro (1961) pointed to the 'inexistence of any legislation which orders work relationships' as 'largely responsible for the insecurity of farmers and peasants', while Velho, in his excellent monograph on Marabá (Velho 1972) described how the nomadic population of cashew nut gatherers could only be controlled by physical coercion. He further observed that this coercion was carefully exercised while the leasing of the cashew nut groves was relatively stable through time, but once the leasing was subject to annual renewal by political decision, and effective possession uncertain, then economic plunder and violent control increased dramatically. This reinforces an interpretation of the violence as inherent in the labour relations of the initial stages of expansion, where increased violence accompanies the rising rhythm of exploitation on the frontier.

During these initial stages, goods leave the frontier regions to enter a market, or a productive process elsewhere. Little of the value created by this economic activity remains within the region, and, in particular, there is little economic interest in investing in the land itself. As already observed, the expansion of extractive activity, for instance, promotes an uncertain occupation of the land, rather than a firm pattern of land tenure. Nevertheless, the expansion of the 'pre-capitalist' economy creates the conditions for its own substitution, both by removing forest cover from the land and by attracting peasant migrants to the frontier. These peasants provide the labour for clearing the land, and wherever they are able they occupy the land, and begin to produce for a market. The land takes on 'value' and an imperfect market for land appears. There occurs a gradual but generic change from the simple depredation of the environment to a developing agricultural economy, and a complex exploitation of labour through the production and market relationships of the national society. In other words, economic expansion and integration into the national economy completes the transition from 'pre-capitalist' to capitalist production on the frontier. But this transition does not go uncontested.

The nature of the contest is revealed in the divergent economic interests of the different fractions of capital operating on the frontier. As suggested previously (Chapters 2, 5, 6) extractive enterprise tends to enter into conflict with enterprise engaged in marketing and credit operations; while speculation in land, in general, is evidently antagonistic to the aspirations of bona fide colonisation companies. In the case of the west of Paraná, once again, it is clear that Lupion's group looked for quick returns from extraction, and through speculatory sales of large stretches of land, while that of the Dalcanales favoured the slow fortune which would accrue from colonisation, and the commercial and credit networks which evolve alongside it. In fact, Lupion had switched his operations from the north to the west of the state precisely at the moment when the spreading commercial network in the north presented too much resistance to his profiteering. This 'resistance' indicates that these fractions of capital do not only differ in their 'preferred' sources of profit but also, and this is the point, repose upon different social relations of production. The 'contest' between them reflects the contradiction between different modes of production.

This paradigm comprehends the struggles on the ground between extractive and speculatory enterprise and its labour force, on the one hand, and, on the other, both the pioneer peasants who strive to stake a claim to the land, and capitalist enterprises, whether productive or commercial. But it is not complete. In the first place, it does not include the increasing antagonism between capitalist enterprise and the peasants in the final stage of frontier expansion (which again reflects a contradiction between different modes of production); in the second place, it is confined to the social tensions deriving from social relations of production in contradiction, and not from these antagonistic relations of appropriation themselves. This latter antagonism looks back to the violent coercion of labour in the 'non-' and 'pre-capitalist' stages, and forward to the primary contradiction between wage-labour and capital in the final stage. The other struggles can all be subsumed within the process of transition to a predominantly capitalist production on the frontier, insofar as this transition requires a regime of private property and promotes a concentration of property in land. A set pattern of land tenure restricts accumulation in the first stages, but is essential to accumulation in the final stage. The key to the violence generated by the transition from one stage to another is the implantation of capitalist property relations.

This conflicted implantation of property relations which characterises the transition is the real process which gives rise to 'moral' interpretations of the violence as the responsibility of unscrupulous entrepreneurs, who practice speculation and subvert the rule of law. In general, such interpretations depart from caricatured accounts of the Brazilian political process as accommodating and essentially 'non-antagonistic' (Schmitter 1971), and describe these frontier entrepreneurs as not conforming to the capitalist and conciliatory norms of the national society, which thereby fail to inhibit acts of violence on the frontier. Without entering the debate on the 'nature' of the national political process, it is not difficult to accept that conflicts in the national society are curtailed by its array of institutional sanctions and display of normative values, and that these sanctions and values are less effective on the frontier. But the logical explanation of the violence which should follow from this is not found in the 'nonconformity' of certain entrepreneurs, but, in the first place, in the fierce political competition between all economic enterprises on the frontier, both for factors of production like land and for the proceeds of accumulation, and, in the second place, in the failure to conciliate the peasant masses who suffer the social consequences of the competition. Any expedient compromises which do occur seem to serve the overall preservation of economic enterprise at the expense of the peasantry.

The 'moral' interpretation of the violence would clearly expect the implantation of private property to have the institutional backing of the law, which should accompany, confirm and ratify the new regime: the legal institution of private property and the social sanctions it commands are essential in the long term to the political control of the means of production. But the political analysis of frontier expansion (Chapters 4 to 7) has revealed that land on the frontier is nearly always titled before it is occupied (or, at least, is titled faster than it is occupied), and, moreover, is titled many times over, so that title provides no security of ownership. This prevailing legal confusion is due to 'dual authority' in the frontier regions and to the intricacies of their legal history, on the one hand, and, on the other, to the speculation and 'bureaucratic entrepreneurship' which are encouraged by the transition from the 'pre-capitalist' to the 'capitalist' stage, and the consequent rise in land prices. Thus, in political terms, while the transition may tend towards a legally secure regime of private property in the long term, in the short term the transition is marked by a generalised lack of legal resolution and lack of security of tenure.

The social consequences of this lack of security are already known. Legal confusion breeds legal disputes and active conflicts which severely prejudice any attempt to supervise the settlement of the land. Different companies, state departments, Federal agencies gain legal rights to different areas of land at different times, but control of land is never more than provisional: State and private colonisations fail; settlement is spontaneous and disordered; State agencies prove 'incapable' of bringing order to the occupation of the land. Everything is in flux. But the overall lack of any apparent resolution carries a very real result. Peasants are continually being forced off the land they farm, and pushed further into the interior and onto the next frontier. What does this mean?

Primitive accumulation

While conflicts continue on the frontier it evidently appears important in the short term which of the different title holders – whether individual, company, state department or State agency – gain control of the land in question. But these appearances add up to more than the sum of their particular parts. In the first place it must be recalled that the control achieved by successful assertion of ownership of the land is never more than provisional; in the second place, very significantly, title holders with very few exceptions only assert their legal rights after the land has begun to produce, or after land prices have begun to rise (because other frontier land is already producing). In other words, they try to achieve economic and legal control of the means of production, the land, once that land is in production, or once its price reflects its future production. Insofar as they are successful in divorcing the direct producers, the peasant farmers, from the ownership or possession of the means of production (separating the ownership of labour-power from control over the means of production) the process of accumulation which takes place is a primitive accumulation. In the particular process which occurs on the frontier, violence plays an important role in the capture of the surplus, whether it is in the extraction of a series of forced levies or forcible capital gains from the peasantry (who have to pay many times over for the land they farm to different title holders), or whether it is in the expulsion of the peasantry from the land so that it may be subject to further speculation, or simply monopolised. In general, primitive accumulation has traditionally been characterised by widespread, sometimes systematic, violence: 'the methods of primitive accumulation are anything but idyllic . . . in actual history it is notorious that conquest, enslavement, robbery, murder, briefly FORCE, play the great part' (Marx 1970).

Any process of accumulation requires the capture of a surplus, and violence serves this end in the process of primitive accumulation. As Velho said of Marabá, 'the regimes of violence and brutal spoliation . . . were economically explicable, looked at from the point of view of the dominant groups, as mechanisms for the extortion of the total surplus' (Velho 1972). This violent exploitation is itself in no way capitalist, but is achieved precisely in the absence of institutionalised capitalist social relations of exploitation. In the national society, wage relations, the creation of surplus-value, and sophisticated price mechanisms assure the capture of the surplus; on the frontier, violence operates in their stead in the short and medium terms. Where capitalist social relations of production are institutionalised, and, where the relations between capitalism and subordinate modes of production are established, then violence will be used to maintain those relationships of exploitation; on the frontier violence itself achieves the exploitation, and may finally be instrumental in imposing institutionalised production relationships. The difference is not only one of degree but of kind.

As long as legal rights to land are still disputed, primitive accumulation can continue. The forced levies on the peasants, and their divorce from the land, demonstrate how legal delays complement the violence, while multiple titling and the inertia of the land agencies illustrates the complicity of local state and Federal State administrations in this pattern of accumulation. Thus 'institutional' interpretations of the causes of the violence are correct – if they successfully locate the contribution of bureaucracy and law in the process of accumulation. Other political apparatuses of the State also make their contribution. Both police and army can and do operate 'against' the law on the frontier, insofar as they exercise violence against peasants and *posseiros* who had legal right (if not necessarily ownership) to occupy the land. But this is not a reversal of role, so much as an extension into frontier conditions of their essential role of guaranteeing and promoting the process of accumulation in the society, and in this case the process of primitive accumulation. Once again, this political promotion of the processs is typical of the mechanisms of primitive accumulation, 'all of which employ the power of the State, the concentrated and organised force of society' (Marx 1970).

The violence which scours the peasants from the land and carries forward the accumulation is possible during the course of the frontier cycle, but becomes difficult to exercise within an emerging regime of private property in the final stage of expansion. In other words, the scope for violence is sharply reduced as the frontier becomes assimilated to the institutionalised structures of the national society, whether at the level of production or of the market. But accumulation continues within these new production and market relationships. In the case of capitalist enterprise accumulation now takes place through the extraction of surplus-value; in the case of petty commodity producers – the small peasants – through market relations which effectively extend the process of primitive accumulation.

The resistance of petty commodity producers to the continuing exercise of violence in the final stage is evident from a last look at the operations of CITLA in the south-west of Paraná. The land had begun to produce in quantity by the time Lupion's gun-men moved into the region, and the stability of peasant society was buttressed by the cultural homogeneity of the peasants and their close organisation within Catholic 'chapels'. Yet the company exerted violent pressures on the peasants to force payments from them or push them off the land – a pure case of primitive accumulation; it plundered the region as if it were still in the 'pre-capitalist' stage and refused to recognise the new production and market relations already in existence. The result was the revolt which swept the region in 1957, and put paid to the company's operations. Of course, the price was high. Hundreds of peasants fled the region, and many others died. New waves of migrants broke upon the region from the south, seeking to claim their own piece of land. The region was submerged. But the process of primitive accumulation did not stop there.

Primitive accumulation: the frontier variation

It must be remembered that in its original formulation the concept of primitive accumulation implied the transformation of small peasants into wage labourers (and their means of subsistence and labour into the 'material elements' of capital). In other words, it referred to the creation of a proletariat through the divorce of the direct producer from the means of production. This transformation of the peasantry provided the historical conditions for the emergence of capitalism as a mode of production. 'The historic process is not the result of capital but its prerequisite. By means of this process the capitalist then inserts himself as a {historical} middle-man between landed property, or between any kind of property, and labour' (Marx 1964). But such a total transformation cannot occur on the frontier, where there is always more land to take into production. While some peasants who are forced from the land they farm may migrate to the cities and there merge with the already swollen marginal populations, and while some may effectively become rural proletarians, working for wages in new capitalist enterprises in the countryside, like cattle-raising, the majority will move onto the next frontier, and repeat the process of a 'spontaneous' occupation of virgin land. Hence primitive accumulation on the frontier is not a historic phase in the evolution of capitalism, but a hybrid mode of accumulation which is clearly subordinate to capitalism: it may impel a mutation to capitalist social relations of production, but may equally expand the sub-capitalist economic environment.

The peasantry cannot be completely 'transformed' on the frontier, because it can reproduce itself by pushing the frontier further forwards. Moreover, the peasantry, at least in the short term, may stay on the land. The purposes of primitive accumulation are served while the surplus is exacted in a series of capital levies. But this surplus must first be created, by application of labour to land, and production for a market. This it is which raises the prices for land. As the peasantry becomes integrated into the market by a complex of commercial relationships, the surplus it produces can now be appropriated more effectively by the price mechanism. The notoriously oligopsonistic marketing structures, inside and outside the frontier regions, reinforced by credit links, will drive prices to the producer down until the peasantry realises negative net returns for its labour. The only way it can continue to survive in these market conditions is by leaching the fertility from the soil (the peasantry is at the end of the 'chain of exploitation' and this is the only way it can pass the burden on). In other words, exploitation through the market leads to a 'decapitalisation' of the farm in the form of the depleted fertility of the soil. This depletion feeds the process of primitive accumulation.

Wherever possible primitive accumulation appropriates the products of nature and transforms them into capital. The primitive accumulation on the frontier appropriates the very wealth of the soil by going beyond the appropriation of that *surplus* created by the application of labour to land, to the appropriation of that portion of value destined for subsistence – to the reproduction of the peasant family. This can occur only in conditions where the peasantry can reproduce itself through taking more land into production – on the next frontier. Thus the accumulation of agricultural capital on the frontier may be conceived of as the appropriation of a surplus created through the continuing combination of labour with land. While labour in general may be easily reproduced, this is not, of course, the case with the land, which is exhausted (it 'depreciates'), and must be replaced by bringing more land into production. The cycle of frontier expansion and the rhythm of accumulation depends precisely on this constantly renewed combination, labour–land. The faster this process can be induced, the higher the rate of accumulation, and, unstable settlement, insecure tenure, and legal delays encourage the exhaustion of the soil and so accelerate the accumulation. If the soil is exhausted the peasant must move on. In this way the divorce of the direct producer from the means of production may be achieved either directly by violence, or indirectly by the production and market relations created by this violence and the 'complementary' institutions of the State. In this way the *individual* peasant may be on a plot of land he considers 'his', and producing 'for himself', but as a *class* the peasantry is effectively divorced from the means of production *through time* – which is the time it takes for the frontier cycle to run its course of accumulation. New land is continually cleared, taken into production, exhausted.

The surplus appropriated within a process of primitive accumulation is transferred from the frontier regions to the industrial and financial centres of the national economy. It is now clear that this transfer may be achieved in a number of different ways, and this is especially true during that part of the cycle when land prices are rapidly rising. Firstly, there is the direct appropriation of the natural environment, and the transfer of 'natural' products to markets or production processes outside the region. Secondly, there may occur more or less direct transfers through the forcible capital gains on land exacted from the small peasants, or the speculative gains on land made possible by their labour. Thirdly, the high productivity of the frontier farmer producing on highly fertile soils, and the extremely low farm-gate prices (imposed from the top down onto the small peasants) open up wide marketing margins and a high rate of commercial profit. Finally, the super-exploitation of the small peasants, involving not merely a transfer of a *surplus*-product, but also the transfer of the *necessary*-product (the products respectively of surplus-labour and necessary-labour), not only raises again the rate of transfer from the frontier, but contributes to reproduce the whole frontier cycle – the complete cycle of accumulation. In these different ways accumulation on the frontier feeds accumulation within the national economy.

The regression to minifundio and latifundio

The violence of the frontier has been explained as an instrument of accumulation. This accumulation is one process which includes the substitution of the 'pre-capitalist' by the 'capitalist' economy, and the de-capitalisation of the petty commodity producers. The violence contributes to complete the cycle of accumulation and to reproduce it on the next frontier. The conflicts and legal confusion spawned by the process of substitution, and the rise in land prices, create the conditions for legal chicanery, political corruption and economic speculation, with their predictably violent results. In the process of decapitalisation, while a predatory agriculture may be economically rational as long as there exists an abundance of land to farm, this rationality needs reinforcing by the threat or reality of a divorce from the land – and violence is again the result. The social insecurity and tension generated in the frontier regions, and the violent expropriation of their peasantry, provoke the disintegration of the pattern of land tenure imposed spontaneously by the frontier peasants, and its regression to the dual regime of minifundio–latifundio. When the legal conflicts are finally resolved, the regime of private property established, and the relations of production on the frontier assimilated and adjusted to those of the national economy, what remains in the majority of cases are latifundio and minifundio (cf. Chapters 2, 5 and 7). Violence is the catalyst both in the perpetuation of the accumulation cycle, and in the reproduction of the classic concentration of land-holding in the countryside.

Quite to the contrary of assertions which are often made of the irrational or 'anti-economic' nature of minifundio, it is the near ideal vehicle for the generation and capture of the agricultural surplus. It carries the accumulation within a continuing combination of labour and land in the cyclical expansion of the frontier. By definition it is cheap, and requires no capital, no new technology, no administration. All it requires is labour and land, and both are abundant. Indeed it is the mass of migrants on the frontier which drive the motor for the reproduction of the accumulation cycle. It provides staple foods for the towns and cities, and, once it is enmeshed by the institutional relations of exchange, must produce uniquely for the market, and so at the same time consume the products of capitalist industry. Finally, the many disadvantages of minifundio – the atomised supply, the lack of capital, the lack of information on prices to the consumer, the lack of credit – in fact favour the process of accumulation of which it is a part. The disadvantages add up to a degree of exploitation which is surely as predatory as the exploitation of the environment in the extractive economy. Exploitation is exercised as if the minifundio were composed only of its natural base, the land, from which wealth can be simply extracted. As Antonio Candido demonstrated so decisively (Candido 1964) this exploitation destroys peasant industry, peasant culture, peasant health, and the basis of the peasants' livelihood, the land.

Judging by criteria of the maximum accumulation with the

minimum of investment and delay, minifundio on the frontier is a successful form of agricultural expansion. Adepts of neo-classical equilibrium analysis are of course at a loss to explain why peasants on minifundio will apply labour beyond the point where the marginal product is inferior to the market wage rate (and indeed to the cost of subsistence), but this is because they assume perfect markets for both land and labour (not to mention goods) when in fact both markets are highly imperfect. Employment opportunities in the capitalist sectors of agriculture are extremely limited, and, except in São Paulo and some regions of the Centre-South, do not pay a 'living wage'; access to land is similarly restricted, and, as Cline (1970) has demonstrated, the price per hectare for identical quality land is far higher for small plots than large estates in Brazil as a whole. Of course such 'imperfect markets' do not account for all the reasons why the peasantry will labour 'irrationally': peasants will usually be in debt to the local middleman who has furnished credit at exorbitant rates; they will labour until the next harvest to pay the debts, buy the medicine they need. But the idea of 'imperfect markets for factors' does suggest, correctly, that the economy of minifundio cannot be analysed except in relation to its complement, latifundio.

Latifundio fulfils the function of 'immobilising' the principal factors of production in the countryside, land and labour. It immobilises land by monopolising it, and it immobilises labour by different forms of extra-economic coercion. The coercion was originally calculated to bind the peasantry to the land of the landlord, and prevent the spontaneous combination of peasant labour and land outside the latifundio. Were this coercion not effective 'the difficulty would be to obtain combined labour at any price' (Marx 1970), and the relations of production within the institution (whether slave labour, indentured labour, debt peonage, or forms of share-cropping) are less important to its existence than the political control over labour that it exercises. Of course this peasantry must have been divorced from direct possession of the land, and now 'possess' it again under conditions which guarantee the capture of the surplus by the landlord. This institution proves resilient because it is the landlord himself who decides what is necessary – and what is surplus – labour, and, in unfavourable market conditions, the peasantry can revert to subsistence. But this does not damage the argument that this labour is immobilised in order that it may be exploited. On the other hand, more and more land is monopolised to prevent its private appropriation by a 'free' labour force. Thus latifundio, historically and contemporarily, has brought the greatest possible quantity of land and labour under its control, not only to use as factors in its own agricultural production (which is often specifically not the case with the land) but to prevent the *alternative use* of these factors (Palmeira 1970).

Latifundio is therefore not capitalist in its organisation but contains a 'non-capitalist environment' which in the contemporary context is subordinate to the dominant capitalist mode of production. This dynamic creation of a non-capitalist environment has taken place by intense political competition for control of land and labour, which was promoted first by the exigencies of commercial capital, then by capitalist production relations in the national economy. Through monopoly of land and extra-economic coercion, latifundio achieves a captive supply of cheap labour and consequently ensures supply of cheap staples and raw materials to workers and industries in capitalism. If contemporarily the institutionalised violence (extra economic coercion) against labour has diminished (which is anyway doubtful on the evidence from Amazônia), this would be in response to an increasing monopoly of land throughout the national territory, and the chronic labour surplus in the economy ('surplus' being a relative concept gauged against the effective monopolisation of the land, prevailing technology, and demographic growth).

It is important to establish that latifundio is in origin and function an institution of direct labour control, and by definition non-capitalist. This is argued by Moacir Palmeira in his exhaustive study of the institution (Palmeira 1970). However he further argues that labour control by extra-economic coercion 'improves the possibilities for primitive accumulation', and the argument is defended on the grounds that latifundio institutionalises the conditions of the historic accumulation of capital by primitive accumulation, which are the divorce of the direct producers from the means of production, and in some way holds them in historical 'suspension'. Of course, primitive accumulation can take other forms than that of the expropriation of the peasantry, such as the plundering of the peasantry by landlords, or its heavy taxation by the State (Preobrazensky 1965), and a notion of a continuing crystallisation of the conditions of this accumulation in the present at least establishes its importance to contemporary capitalist growth. But the notion as applied to latifundio, even if it establishes the capture of a surplus, fails to analyse adequately how labour can be 'divorced' from the land, but kept on it at the same time (and also fails to specify the class relations of exploitation). Discussion of these questions must wait until the following chapter. For the moment, it is sufficient to note that traditionally accumulation in the countryside, whether in minifundio or latifundio, has advanced through the original divorce of the direct producers from the land, and their 'repossession' of it in conditions which guarantee the capture of the surplus; and that, traditionally again, the phenomenon of the pioneer frontier represents the discrete attempts of millions of individual peasants to escape the 'conditions of accumulation' in the countryside.

Here the argument has come full circle. The process of primitive accumulation on the frontier impels the regression to the dominant duality in the countryside, latifundio–minifundio, in a dynamic reproduction of the highly concentrated system of land-holding in Brazil, which itself provides the *general conditions* for the expansion of the pioneer frontiers, and the appropriation of a surplus through primitive accumulation. In this way it can be seen that frontier expansion has its own cycle which is in some sense 'self-perpetuating', but in no sense 'independent' or isolated. It derives from the extreme monopoly of land in latifundio, and the extreme pressure on land in minifundio, and the two institutions are far more inter-dependent than is commonly understood – both serving the same ends as integral parts of the same political economy. Just as political competition for land originally established this duality, so an intense political struggle for land reproduces it on the frontier. This struggle is violent and the violence an instrument of primitive accumulation, which, in its turn, is 'self-perpetuating' to the degree that expropriation of the peasantry on the frontier propels the frontier forward, but which can only continue in subordination to the dominant capitalism of the national economy.

The state and violence on the frontier

A partial and descriptive definition of State which specifies its relation to violence in the society might be that of a centralised organisation which plausibly claims to monopolise ultimate control over the use of force within a given territory. Indeed, it is generally agreed that of all the standard, 'Weberian' State-building processes, such as the elimination of rivals, the formation of coalitions, the extension of protection and the routinised extraction of resources, it is the consolidation of control over the use of force which is the most important. Empirically, an inverse relationship is usually observed between the extent and acceptance of private violence and the level of State control over the means of coercion. At first sight, all this seems in accord with the use of force in the process of primitive accumulation. It is the intervention of the State, and the application of institutionalised violence which defines the process: the mechanisms of primitive accumulation 'all employ the power of the State, the concentrated and organised force of society . . . force is itself an economic power' (Marx

1970). The police and army are present on the frontier and engage in armed operations. They carry out court orders, and implement administrative decisions; they clear peasants from the land, protect the boundaries of big estates, and back up the surveyors of the big companies. In extreme cases, they may quell revolts, and discipline and punish a recalcitrant peasantry. But they do not command a monopoly of violence on the frontier, although they may claim it as their prerogative. On the contrary, while this institutionalised violence exists it is spotty and sporadic in comparison with the exercise of private violence, which is widespread and endemic. Similarly, the performance of the bureaucratic and legal apparatuses is seen to complement the violence, but not so much by legitimising and organising coercion by the State as by tolerating, promoting, sometimes precipitating, the use of private violence. The necessary intervention of the State in the process of primitive accumulation explains the use of public apparatuses for private purposes, but it cannot explain the extensive use of private violence which, in principle, runs contrary to the central reason of State.

In fact, the paradox presented here is only apparent. The reason of State is not only political but economic. It exists not in isolation but in relation to dominant classes in the national society, and in relation to a world economic system. In this sense the political role of the Brazilian State is largely determined by its existence as a dependent capitalist State. Its primary political function is the protection and reproduction of the dominance in the society of the capitalist mode of production, and, as such, one of its paramount political tasks is to promote the appropriation and concentration of the surplus created in the processes of accumulation which are subordinate to capitalism. The implications of the larger role will be investigated in the final chapters, but it is important to note now that this particular task is capable of distorting the traditional 'manifestations' of the State, such that it is not only tolerates, but welcomes 'private' violence insofar as it catalyses the appropriation of value.

But the paradox finally collapses not from the weight of 'grand theory' but in the face of arguments already advanced regarding the direct representation of economic interest within the State administration, and the direct links of the State itself with large enterprise. The distinction between public and private is, anyway, a legal one – that is, it is defined by the State in the first place – and this distinction becomes increasingly blurred at the level of accumulation. It has been suggested, for example, that the inertia of the State's bureaucracy on the frontier has in some sense 'permitted' the use of private violence. But, in the present perspective, this same bureaucratic inertia masks an intense political activity which is specifically directed to guarantee the continuation of the conditions of accumulation in the countryside, and, in the case of primitive accumulation, to advance it by violence. In general, these conditions can only be maintained by repression in different forms; in the particular case of the frontier, the political imposition of the primitive accumulation can only be achieved by the open use of violence. As Florestan Fernandes said some ten years ago (Fernandes 1968):

> In essence, the political inertia is more apparent than real. Apathy and passive acceptance exist amongst those 'condemned' by the system, the results of collective demoralisation, and an ultra-repressive transitional regime. But those who direct this regime, in the countryside, in the towns, and in positions of political, military, and legal control of the State apparatus, are in constant political activity, effervescent and efficient . . . this activity, as much at the socio-economic and cultural levels, as at the political level, is hidden and almost invisible, especially in what refers to the urban sectors and those who work through the State apparatus. It is concealed behind stated objectives like 'the preservation of social peace' or the 'acceleration of economic development'. The same does not occur (and could not occur) with the more or less privileged sectors of the agrarian economies (and their agrarian–commercial and agrarian–industrial links). The 'defence of order', for these sectors, involves the open unmasking and the undisclosed use of violence (even when it is explained as having in view the 'benefit' of those who do not understand what they are doing). This happens because these sectors cannot merely manipulate their means of indirect social control or repression – staying within the limits of the protection, pure and simple, of their socio-economic, cultural and political privileges. They have to go beyond this, because they must prevent the pre- or sub-capitalist infrastructure from collapsing, and so destroying, the material base of the type of capitalist accumulation they carry out. For them, therefore, the situation is one of permanent conflict and political struggle, even though the conflict and struggle are negative (preventing the dispossessed and poor populations in the countryside from achieving conditions from which they might impose genuinely capitalist forms of the labour and product markets in the agrarian economies).

The State and accumulation in the contemporary context

Fernandes' observations find ample confirmation in the contemporary context of the Amazon region where the State has widened the gamut of its intervention in recent years to extend the scope and accelerate the speed of accumulation in the countryside. The SUDAM incentives, designed to reap the maximum return for private capital with the minimum investment, attest to the primacy of accumulation (and not 'development'): SUDAM's first development plan (First Five Year Development Plan 1967–71) appeared well after the fiscal incentive scheme had come into operation. This accumulation was carried forward by the increased scope for investment implied in the emphasis on an extensive and 'entrepreneurial' occupation of the Amazon, mainly by cattle ranches and mining concessions. As growth in agriculture has always been achieved through an expansion of farm area (Chapter 3), the State is pursuing a traditional pattern of growth: what is different is State support and promotion of the general conditions of accumulation, which include monopoly of land. This refers not only to the emphasis on cattle – implying huge estates with minimal absorption of labour after the initial phase of implantation – but also to the central role of SUDAM again, and later INCRA, in securing the private, large-scale, and speculative appropriation of land: SUDAM 'unofficially' helped instal some 332 estates of an average 50,000 ha which had presented no development project. State-promoted monopoly of land means the expulsion of small owners, and 'token' production, while owners await the rising of land prices. Finally, not only has nothing been done to contain the massive migration to the new Amazon frontier, but this migration has been actively impelled by the State, and especially INCRA, through its demagogic 'impact-projects', colonisation schemes, and the construction of the Transamazônica. The result of the 'social colonisation' has been to foster a captive labour force, which is susceptible to terrible exploitation by the large enterprises. The compassion of the State for the plight of the peasants of the North-East has predictably raised the profits of private enterprise.

The widened gamut of intervention introduces a greater range of production relations on the frontier. There are the frontier peasants who clear the land, at no cost to capital, and are then expelled from it; capital is left free for land speculation elsewhere, and the peasants impelled to clear more land. There are also the *volantes*, composed of dispossessed peasants or 'seasonal' labour from the North-East, who are transported to the frontier to clear forest for cattle ranches or bring in the harvest, and then laid off. Without resources they must enter the 'agricultural reserve army' of labour. Some of the deprived peasants and abandoned *volantes* may find permanent employment on cattle ranches or in other capitalist enterprises, but this does not necessarily mean that they are 'rural proletarians'. They are the *peões*, and may be integrated into production relations which more resemble slavery or villeinage.

In this way the advance of capitalism tends to retain and transform traditional work relations. The *aviamento* system continues through different mutations, and the collective hiring of labour gangs locally or in the North-East may include similar debt relations. Not only are labour laws not obeyed, but within the system of 'contracts', which is *aviamento* and its modern mutations, peasants and workers are on occasions murdered for the money owing to them by the boss or 'contractor'. Even massive capitalist enterprise in Amazônia like Volkswagen or the JARI estate have exploited their workforces through relations of production which are clearly sub-capitalist: President Medici's horror at the inhuman work conditions on this latter estate provoked the reply that 'contractors', and not the company, were responsible for three quarters of the 5000 workers. Therefore, although it is possible to find the incipient formation of an agricultural proletariat in select areas of the new frontiers, in general this exploited population is in no way 'free' to sell its labour-power where it will. Capitalist penetration in Amazônia clearly does not bring about a general modernisation of work relations.

Nevertheless, the intervention of the State in monopolising the land, in the migration to the frontier and the creation of a labour reserve, in the clearing of the forest and preparation of the soil, is designed to support accumulation of capital – whether through capitalist or sub-capitalist production relations, or through mixed and transmuted forms of production. The presence of sub-capitalist production relations within the estates of large capitalist enterprises is especially illustrative of the acceleration of accumulation through the intensification of exploitation of the labour force. This exploitation is now backed directly by the State. The State has brought capitalist enterprise deep into the interior, hard onto the furthest frontier, but the process of primitive accumulation remains essentially similar; the exploitation and expropriation of the peasantry in a constantly rising rhythm continues to fuel the accumulation, and the accumulation continues to depend on State intervention, and the extensive use of violence against peasants and workers. The capitalist penetration of Amazônia, and the 'modernisation' of agriculture, changes nothing for the dispossessed populations of the interior, except that they are yet worse off (not enjoying the few guarantees of the traditional paternalistic order, yet not benefiting from rational and modern work relationships). Not only are social inequalities not reduced, but the rate of exploitation is increased to maintain the viability not only of the agrarian but of the industrial economy. The burden is passed onto the peasantry once again. In broad terms the State spearheads the 'capitalist' occupation of Amazônia to resolve the contradictions of its small partner role in international capitalism. Further, as much of the capital now promoting different forms of primitive accumulation in Amazônia is foreign, the State defends its initiative by concepts of 'national security' – which are the ideological counter-balance to the internationalisation of the process of primitive accumulation in Amazônia.

These comments concur again with Fernandes' interpretation (1968) of the 'so-called inaction of the economic, cultural and political elites'. He refers to:

> the interests of these elites in maintaining the status quo, and to the specific interests of the privileged sectors in the countryside, effectively committed to the social reproduction of labour-power which does not succeed in becoming a commodity, or which only becomes an extremely undervalued commodity . . . It is the 'most modern' strata, 'active' and 'influential' in the agrarian economy, who head the crusade against any change whatsoever which might alter the 'structure of the situation', or simply threaten their power of decision and domination . . . What interests them, exclusively, is to eliminate or restrict any rapid and uncontrollable absorption by the agrarian economies of specifically capitalist forms of growth or development, which might weaken or destroy their ability to impose themselves on the 'normal' functioning of the internal market and the modes of production. Paradoxical as it may appear, the 'forces of order', and the 'defence of the social peace' are identified in reality with the indefinite survival of economic, social and political iniquities, which are incompatible with 'mature capitalism'.

But while there is no change in the broad lines of the domination of the sub-capitalist modes (or environment) by the capitalist mode of production, what does change, with the penetration of Amazônia by large capitalist enterprise for instance, is the *form* of this articulation of the modes of production. Similarly, the role of the State in achieving and underpinning this articulation politically, does not change its broad character, but changes in content with the changing form of articulation. These aspects of the frontier experience – economic and political, empirical and theoretical – will be examined in the final chapters.

9

The frontier and the reproduction of authoritarian capitalism

It has been established that the pattern of frontier expansion is not fortuitous, but proceeds in response to larger forces within the national economy. In particular political intervention and violence are seen to promote a specific process of accumulation, which contributes to national economic growth. The relationship between frontier and national economy is interpreted, in purely economic terms, as achieving a transfer of value from one to the other. But the moving frontier does more than merely feed the growth of the national economy through primitive accumulation. Frontier expansion extends the boundaries of this economy, and by its advance creates an economic, from a natural, environment. The frontier experience, the political intervention and the violence, is not merely the effect of specific production and exchange relations, but participates in a complex of such relations in the countryside, and can impose these relations. Therefore, in addition to a particular interpretation of the relationship of the frontier to the national economy, there exists a clear need for a broad conceptual scheme which can successfully locate the place of the frontier in the formation of the national economy – a global theory of frontier expansion.

By definition the frontier exists on the periphery of the economy, and the scheme of 'centre–periphery' relations seems a logical choice for the location of the frontier. But such a scheme is purely descriptive, and in practice is given very different conceptual contents (Balan 1974). In general it is agreed that the 'centre' should represent the central capitalist countries of the world economy, and 'periphery' the underdeveloped economies, and that the economic performance of the periphery – as influenced by changes in import capacity, the terms of trade, or capital flows – depends on that of the centre (Dos Santos 1970). This dependency can be presented as a linked chain of economic exploitation (or a series of metropolis–satellite relationships, Gunder Frank 1967), which descends to the furthest point of the periphery. The economic contradictions of the centre can be contained by directing their effects downwards to the next link in the chain – which in economic terms means increased exploitation (Ianni

1970). In such a model the frontier could be the final 'link' where such contradictions find violent release.

The scheme of 'centre–periphery' relations, both functional and geographical, has been extended to the study of the economic and political structures of particular countries in Latin America. Here the intention is either to investigate the refraction and refinement of 'central' forces by the specific structures of the periphery (Cardoso and Faletto 1969; Sunkel and Paz 1970), or to investigate the reproduction of typically international relationships in national contexts – in the form for instance of 'internal colonialism' (Stavenhagen 1971; Casanova 1970). While such studies contribute more to an understanding of specific realities than 'unilateral declarations of dependency', their view of the periphery often remains captive to the analytic criteria and categories of the centre. The concept of 'internal colonialism' successfully suggests an idea of economic penetration and political encapsulation of the periphery, by larger forces, and even establishes the extraction of a surplus from a 'colonised' peasantry. But such accounts have been correctly criticised for purporting to explain the 'colonies' existence in terms of race and culture: 'racial and ethnic entities are treated abstractly as if their internal class structures are irrelevant to their existence as groups and to their political and ideological practices' (Wolpe 1975). In other words, inside the 'colony' there will be exploiters and exploited, and, in obscuring exploitation by class, the concept of colony tends to transpose primary contradictions to a secondary, and less specific, level.

All these concepts are more or less descriptive and support arguments which tend to relate centre and periphery by a series of analogies and 'correspondences'. In other words the concepts capture the surface of events rather than revealing the underlying reality. In their stead, the centre–periphery relationship can be structured in terms of more theoretical concepts like 'mode of production'. Analysis at this level of abstraction cannot claim to describe the reality, but may suggest structural determinations and contradictions which are capable of 'informing' the reality and giving it meaning. While clearly privileging economic relations the concept of mode of production should not be conceived as implying a direct determination at the economic level of political initiatives and intervention. In general, such concepts can only provide the 'pre-conditions' for studying concrete realities, and immediately demand further determination if they are to explain how contradictions lead to confrontation, or how social forces and institutions which are 'structurally' incompatible come into conflict. For instance, Lenin (1967) discussed the basic contradictions, as he saw them in the Russia of his day, between the 'archaic' institution of the great landed estates, and the 'advanced' form of industry – two productive processes which are defined by different modes of production. Such structural 'contradictions' exist in Brazil today, but lead not to confrontation but to an articulation (linking) of different modes of production in which, as will be seen, the frontier plays a special role. In particular the further determination of the concept in this analysis, will include, specifically, the idea of a 'dominance of the political' which refers to the relative autonomy of the State in relation to the economically dominant class; and to the necessary intervention of the State to promote capitalist growth, and especially to guarantee the political subordination of the peasantry.

The analysis will attempt to locate the place of the frontier in the formation of the national economy. But this cannot be done without first discussing the formation of the peasantry itself, both in the countryside at large and on the frontier. Many of the elements of this aspect of the analysis have already been suggested – not least the 'immobilisation' of the peasantry on the large landed estates (latifundia). This immobilisation will now be theorised in terms of the peasantry's relation to the mode of production in the countryside – but this relation is not directly determined economically. The idea of immobilisation also recalls the repeated observations in general discussions of the peasantry of its 'subaltern' or 'underdog' position and, especially, its 'domination by outsiders' (Shanin 1971). Tepicht (1969) asserts categorically that the peasantry 'has never been represented in any historical formation where it is present by a ruling class. In one way or another it is always in a subordinate position.' So the formation of the peasantry whether in the countryside at large or on the frontier must be analysed not only in relation to the productive system in the countryside, but also in relation to its political subordination. Historically, it is a purely political intervention which has defined and imposed the relation to the mode of production. Finally it will be seen that the peasantry produces within one or more modes of production in the countryside and at the same time is directly subordinate to a landlord class which represents the dominance of *another* mode of production within the social formation as a whole. And it is through the articulation of these different modes of production that the centre–periphery relationship is constructed theoretically. To begin this construction in a contemporary context, it is necessary to adopt a historical perspective.

A historical perspective

The interpretation begins with the transition to capitalism in Europe.

It is during this economic revolution that the peasantry is transformed, and finally 'disappears' for all practical purposes from certain of the central capitalist countries. This was theorised in Marx by the stage of primitive accumulation, which, in its internal aspect, achieves a once-and-for-all divorce of the direct producers from the means of production in the countryside. This divorce is theorised in turn as a necessary condition for the development of capitalism: in conjunction with the institution of landed property it forms the 'double-mill' which drives labour to the cities, and promotes centralisation of capitals (expropriation of the peasantry takes the form of supplementing one kind of private property by another). In this way Marx successfully theorises the transformation of the peasantry, and its relation to capitalist development, as he knew it: 'once capital and its process have come into being, they conquer all production and everywhere bring about and accentuate the separation between labour and property, labour and the objective conditions of labour' (Marx 1964). Capitalism is thus defined as the universalisation of commodity exchange, including, crucially, labour-power itself as a commodity. But as early as 1885 (letter to Zasulich 23 April 1885) Engels suggests that the 'historic fatality' of this transformation may be confined to Western Europe. In other words, he opened the possibility for a revision of the theory. This revision is yet more necessary today, when it is seen that the expansion of capitalism as a world system has in no way precipitated the 'disappearance' of the peasantry in the third world.

This revision demands closer examination of the transition to capitalism, and consideration of the developments which follow the industrial revolution – within a more general genealogy of capitalism itself. First, one or two observations can be made on the passage from feudalism to capitalism. Feudalism in its archetypal form had not existed for centuries prior to the industrial revolution. It experienced a slow transformation, under the aegis of the Absolutist State, which has been defined as the State of the feudal landlords, in conditions of feudal crisis (Anderson 1974). In attempting to protect feudal property relations, this State simultaneously established the conditions for the development of capitalism. This is not to say that capitalism as a mode of production is suddenly dominant. In the period of transition, capital – largely commercial and usury capital – rules, but rules outside the sphere of production (Poulantzas 1968). In fact the transformations in feudalism gave considerable economic scope to that fraction of the bourgeoisie engaged with commercial capital and protected by the State through a system of legal monopolies. Given this scope, the commercial bourgeoisie could practice an external primitive accumulation, during what is known euphemistically as the mercantilist era (Marx 1970; Anderson 1974), while the Absolutist State intervened politically to promote the internal aspect (divorce of the direct producers from the means of production). Both aspects of primitive accumulation contributed to create the conditions for manufacture, and finally for the industrial revolution. Thus when the bourgeois political revolutions occur, they mark the consolidation at the level of the State of a victory already achieved at the level of distribution and production. The transformations in feudalism prepared the way for the original form of capitalist development, and it is this form, as we know from Marx's (1970) discussion of Wakefield, for instance, which demands the divorce of the direct producer from his means of production.

This original development is obviously the result of the transformation of feudalism 'from within'. When divorced from its means of production the peasantry has no option but to migrate to the cities and sell its labour-power. But this impulsion to the cities depends as well on the other stone of the 'double-mill', the institution of private property, which implies the absence of *free land.* Where free land existed then feudal forms of property holding and work relations proved far more resilient. In the presence of an 'open frontier', the Absolutist State had to intervene to immobilise labour politically, usually by the introduction of serfdom. Without such systems of labour control it would have been impossible to extract a surplus from the peasantry. In the Russian case, for instance, we see the introduction of serfdom, and the flight to the frontier of those who can escape (Trotsky 1965). Both serfdom and flight lead to the relative unimportance of towns and cities, and a 'delayed' process of social differentiation. This is not a suitable social and economic environment for the emergence of a strong bourgeoisie: on the contrary, most accumulation continued to be controlled by the State within specific systems of labour control. In other words the conditions were not present for the development of capitalism in its original form, but nevertheless these States will experience the diverse pressures, both military and commercial, deriving from the expansion of capitalism 'at the centre'. These increased from the time of the bourgeois revolutions, constituting a secular process of 'combined and uneven development'. The pressures emanating from this autonomous process of capitalist growth elsewhere do not lead to the dissolution of the systems of labour control; on the contrary, these systems may be reinforced and extended in order to produce more for the growing 'bourgeois' markets (in the Russian case, again, this was Engels' 'second serfdom', when the 'capitalist era in the countryside is ushered in by a period of large-scale agriculture on the basis of serf-labour services', Engels 1928). The peasantry remains on the land, and remains politically immobilised, while first commercial capital, and later capitalist social relations of production, begin to dominate the economy. The development which takes place is necessarily different from the original form of capitalism (with its universal creation of 'free labour') and can be called ***authoritarian*** capitalism (Velho 1976). This form of capitalism is 'reflexive', and develops out of systems of labour control existing prior to the transition to capitalism at the centre and its expansion from the centre.

It was said that the bourgeois revolutions were the consolidation at the level of the State of a revolution already achieved at the level of production. The bourgeoisie in this case is both economically dominant and politically directing, in Gramscian terminology. In other words it is hegemonic, and it is not hard to see why classical Marxian analysis conceived of the capitalist State as an 'instrument of the bourgeoisie' (Lenin 1972). The bourgeoisie secures the State in order to guarantee the reproduction of capitalist social relations of production. In authoritarian capitalism, on the contrary, it is the State which oversees the progress of the bourgeoisie. The State develops fast in order to exercise the coercion necessary to exploit the peasantry; its growth far outstrips that of 'civil society'. If with the growth and expansion of the capitalist mode of production the bourgeoisie becomes dominant, it is still not directing. In short, the economically dominant class will not be politically dominant (is not hegemonic) and remains generically weak. Very often it must share political power with different classes and class fractions in the power bloc (Poulantzas 1968). It is not the bourgeoisie but the State which achieves the 'bourgeois revolution', and the State enjoys a far greater autonomy of the dominant class or class fraction than in the original form of capitalism (and particularly in moments of crisis). In authoritarian capitalism the transition to the capitalist mode of production resembles a period of 'passive revolution' far more than it does the bourgeois revolutions of Western Europe.

In an emergent authoritarian capitalism, given the presence of free land, the State intervenes to keep labour on the land. This it does through specific institutions of labour control, but the precise internal organisation of the institutions – whether work relations are defined by slave labour, serfdom, indentured labour, or debt peonage – is unimportant, as long as they guarantee the supply of labour on the land. The land itself is of course monopolised to prevent its use by an 'independent' peasantry. It is not surprising in these conditions that the most powerful participant in the power bloc is the land-owning class. Examples of such political power are Russia in the nineteenth century, Brazil and Latin America after independence (a more complex case which requires analysis of the formation of the State after the 'dissolution' of the colonial administration), and South Africa in this century. Elsewhere of course the presence of free land led to quite different developments, as in the United States and Australia – both of them countries which today are noticeably lacking in a peasantry. In these cases no State existed which could impose labour control, and in these conditions it is the frontier – the very opposite of the political immobilisation of the peasantry – which is the formative influence in the growth of the economic system. What is crucial for the argument, therefore, is that the double-mill of expropriation and landed property is as necessary in authoritarian capitalism as in its original form, but in a very different way. In authoritarian capitalism labour is kept on the land. It is politically immobilised so that it may be economically exploited. It is for this reason that this capitalism, in contrast to the original, is characterised by a marked 'dominance of the political'. It is this which finally characterises the difference of authoritarian capitalism which has been widely, if only implicitly, recognised (e.g., see discussion by Samir Amin 1970).

Brazil is a clear case of the emergence of an authoritarian capitalism. Labour is coerced to work within institutions of labour control, in response to the demands, first of commercial capital in the world market, and, much later, to the pressures generated inside the economy by capitalist social relations of production. The dominant institution changes as new products find a demand in the international market, and as minerals are discovered, but these different units of production – the plantation and the *engenho* (slave labour), the gold mines of Minas Gerais (slave labour), the cattle fazenda (servitude and debt peonage), the early coffee fazenda (slavery and debt peonage) – are all simultaneously institutions of control, and are authoritarian by origin and function. In them economic and political power are fused, and therefore, by the beginning of the Empire, political power is local and regional, and clearly associated with the ownership of land. This fact was reflected both in the property qualifications for the restricted number of 'electors', and, more importantly, in the *coronel* system (both formal and informal), whereby, as the name implies, political and military power rested with the landowners. In short, historically, the social and economic basis of Brazil's development is the total immobilisation of labour.

At first sight, this historical basis of Brazil's economic development appears incompatible with the dynamism of the pioneer frontier as described in previous chapters. But this is to ignore the crucial changes in the historical conditions of accumulation in Brazil, which occurred early in the present century, and were complete by the thirties. Prior to this time labour was scarce and land far from being monopolised; the immobilisation of labour within institutions of labour control was essential to accumulation in the countryside. After this time, as suggested in the discussion of Normano in Chapter 3, labour became increasingly abundant in the economy as a whole, due to immigration and secular demographic growth, and this 'abundance' became acute in certain regions of the countryside, such as the North-East, where more and more land was monopolised. As long as relations of exploitation in the countryside remain unchanged, however, this did not inhibit a continuing immobilisation of labour on large landed estates. But there was now a labour 'surplus' over and above the immediate needs of accumulation, which could migrate to the cities or the frontiers. This idea of labour 'surplus' has little meaning outside of specific social relations of production and property relations, and it was precisely the extended monopoly of land, denying this 'surplus' access to it, which created the general conditions in the countryside for the advance of the pioneer frontier.

The advance of the frontier creates the conditions for a continuing process of primitive accumulation, and on the frontier itself the process is correctly theorised in this way. But, as suggested in the last chapter, it is insufficient to theorise the 'extensive' development which took place in Brazil and other authoritarian capitalist countries simply as a primitive accumulation. It is not that the different elements of primitive accumulation are not present in the plunder of nature, the practice of a predatory commerce and even in the transformation of the economic environment through an expropriation of the peasantry. Neither is it incorrect to suggest that these countries could not compete in mercantilist or colonial adventures in order to accumulate, and so did so 'primitively' within their own boundaries. But the important point is that the process of primitive accumulation could not create the conditions for a 'complete' transition to capitalism, precisely because this accumulation occurred in response to capitalist growth at the centre. On the contrary, the process led to the extension of the institutions of labour control, and the perpetuation of an economic environment which was clearly not capitalist. Within this environment labour was exploited not through the wage relation, but through a combination of extra-economic coercion and monopoly in institutions of labour control. This monopoly immobilised the peasants by suppressing alternatives for economic survival: even in the contemporary context, where a surplus of labour often makes the peasants 'free' to move, they stay on the land wherever possible, in the face of a generalised lack of alternatives. This 'immobilization' and 'lack of alternatives' do not simply indicate a process of primitive accumulation, but rather express social relations of production which are *subordinate* to capitalism. The 'double mill' operates not in the transition to capitalism, but within *sub-capitalist* modes of production.

The articulation of the modes of production

The internal laws of motion of capitalism are everywhere the same, and the differences between the original form of capitalism and authoritarian capitalism do not denote different modes of production but different *economic systems*. What distinguishes authoritarian capitalism as an economic system from the original form is the articulation of the capitalist mode with other subordinate modes, and the continuing importance of this articulation to the expanded reproduction of capitalism in the 'authoritarian' social formation. The development of capitalism at the centre leads to the almost total dissolution of these sub-capitalist modes, but in the periphery the integration into the world market, and subsequent commercial pressures, led on the contrary to a reinforcement of sub-capitalist modes, as increased exactions were made on the peasantry, and slaves, to supply the new markets. At least during the period of competitive capitalism at the centre, the surplus produced in sub-capitalist modes with a high proportion of living to 'dead' labour (relative availability of factors) counteracted the tendency for the rate of profit to fall (Marx 1974), and the articulation of different modes was instrumental in maintaining the average rate of profit in the world capitalist system (Laclau 1969). Lenin in his essay on Imperialism (1970) foresaw the export of capital 'expanding and deepening the further development of capitalism throughout the world', but even with the transition to monopoly capitalism at the centre and the implantation of industrial capitalist modes in peripheral social formations, the sub-capitalist modes were not dissolved, but further reinforced – so demonstrating the intimate interconnections of the modern and 'feudal' sectors, and the contribution of the sub-capitalist surplus not only to the reproduction of particular non-capitalist modes but to the reproduction of the system of authoritarian capitalism.

It is recognised that the sub-capitalist modes of production may themselves be differentiated one from the other by their internal structures, and may have a certain but not total capacity for reproducing themselves. But the use of a general concept of sub-capitalist modes can be justified on two counts. In the first place it is used *pro tem* for the purposes of explaining authoritarian capitalism as an economic system. The study of these different modes as separate structures tends to lead to a formal structuralism, which obscures the process which defines them as modes of production – the real appropriation of the surplus. The investigation of specific structures can be safely postponed until the appropriation and transfer of surplus across modes of production, and to the capitalist mode, is established. In short, what concerns the argument is not the separate 'existence' of modes of production but their articulation. Historically the peasantry's surplus was expropriated by different ruling classes in different modes of production, but contemporarily that surplus or part of it is transferred to the capitalist mode; so whatever the differences between these modes they can at least be defined in common by their subordination to capitalism. In the second place there are good a priori and empirical reasons for such an approach, founded in the most important economic development of the last two centuries, which is this same expansion of world capitalism and the penetration of peripheral formations by the dominant capitalist mode. It now appears probable that at this stage of the articulation the 'laws of motion' of the subordinate modes are becoming more and more similar (and the relationship of the peasantry to the means of production nearly everywhere the same). There are no 'pure' modes, if there ever were, but only 'exceptional forms' which have been transmogrified by a continual process of dissolution and reinforcement and vice versa. Appropriation takes place and social relations of production are reproduced in possibly very different ways than would have occurred in 'ideal' conditions, which are the conditions of a mode of production 'existing' in isolation. The emphasis again falls upon the fact of articulation.

The analytical advantages accruing to an emphasis on process, rather than a fixation in structure, are immediately apparent in the case of Brazil, where the pressures of commercial capital (and finally the penetration of capitalist social relations) did not so much transform a pre-capitalist economic environment (which apart from scattered and nomadic tribes of Indians did not exist) but actually created what is necessarily a sub-capitalist environment. It is futile in this case to enter tortuous debates over the actual mode of production in the countryside, and whether for instance it is feudal or not. It is more pertinent to remember Engels' observations that the basic social relations of production are limited in number and are 'invented' and 'reinvented' by men in history, so that different 'forms' of such relations can exist in a variety of historical periods and socio-economic settings (Engels 1928). For instance 'it is certain that serfdom and villeinage are not a specifically medieval form, it occurs everywhere or almost everywhere where conquerors have made the native inhabitants cultivate the soil for them' (to Marx, 22 December 1882). Where such forms are sufficiently consistent and refined then they may constitute, in conjunction with a specific degree of development of the productive forces, a mode of production such as the feudal mode, or the Asiatic mode, and it is important to note that, in history, peasant production and appropriation of peasant surplus, has occurred within different modes. In particular, in nearly every historical social formation, some peasants produce within the petty commodity mode, which everywhere is subordinate to the other modes of the formation. Pioneering peasants are petty commodity producers, and clearly subordinate in the case of Brazil, not only to the dominant capitalism but to the major sub-capitalist mode created by the expansion of capitalism.

The sub-capitalist modes are created, extended and reinforced because the value created within them can later be transferred to the dominant capitalist mode of production. Even before the transition to capitalism at the centre, peasants producing on the periphery within institutions of labour control often produced for a market, and from the end of the sixteenth century this might be the world market. As Frank argues, latifundia, for instance, which are isolated and dedicated to subsistence, may not always have been so, but may have reverted to subsistence following a fall in demand for their products on the world market (Gunder Frank 1967). Historically, landed property in institutions of labour control looked to produce not only use- but also exchange-values.

In the contemporary context, peasantry subsumed within sub-capitalist relations of production produce commodities, or, more strictly, 'goods' which enter the capitalist mode directly as raw materials or indirectly as staples. The transfer of value implied in the production of these 'goods' may take different forms. The goods themselves may be transferred by plunder, exchange of non-equivalents, or similar means. These goods are anyway produced by so-called 'cheap' labour, and when staples, for instance, enter the capitalist mode they lower the cost of reproduction of the labour force in that mode. This labour is 'cheap' only by the criteria of capitalism of course, which benefits from it by enforcing, again, certain conditions of exchange: surplus-labour, in the theory, only has meaning in relation to necessary-labour, and by indirectly pumping out increased surplus-labour from the sub-capitalist modes necessary-labour time in capitalist production can be compressed and so raise the rate of absolute surplus-value. Similarly, if labour moves directly from the sub-capitalist mode (where, in seasonal migration, for instance, at least part of its subsistence is guaranteed) to the capitalist mode, then real wages may be paid at below subsistence, and from the capitalist viewpoint its labour-power may be bought at a cost below its value (which is anyway very low). Not only peasants within institutions of labour control but minifundistas and petty commodity producers in general can be assimilated to this analysis, insofar as by labouring on neighbouring estates or enterprises, or by clearing land on the frontier (inside such work relations as *troca pela forma, comodato, parcería*), they provide labour-power at a cost below its value (either directly to the capitalist mode or to the major sub-capitalist mode).

In general, the empirical emphasis on the production of goods, or the theoretical emphasis on the transfer of value, effectively discount other misconceived interpretations of the articulation of the modes of production. Firstly it is argued that the articulation satisfied a permanent need of capitalism to realise its surplus-value in pre- or sub-capitalist markets. The weakness of this argument can be demonstrated by the traditional methods for imposing and extending such environments, all of which employ coercion and violence: capitalism can certainly force peasants off the land, and can force them to labour on the land, but it cannot force peasants to buy its commodities (Bradby 1975). Historically, it is safe to say, force has not been used to introduce commodity exchange. Secondly it is asserted that the articulation is necessary in order to expand the potential labour force. At some period in the past this assertion may have seemed plausible, but it is almost nonsensical today in a country like Brazil where the cities are swollen with 'marginal populations', the countryside is swarming with landless peasants, and capital 'embarrassed' by a massive labour surplus.

If the arguments for the transfer of value from sub-capitalist modes to the dominant capitalist mode are correct, there seems no reason to believe that the articulation is a transitory phenomenon, but every reason to think of authoritarian capitalism as a distinct genus of development. The peasant product is economically necessary to the reproduction of the social relations of production not only in the sub-capitalist mode but in the authoritarian capitalist formation as a whole. But it has been argued (notably by Rey 1973, and Bettelheim 1972) that the articulation occurs in two stages, beginning with capitalism incorporating the 'goods' produced within the sub-capitalist modes as part of productive capital. During this first stage when commodity exchange is generalised in the world market (this corresponds to the period of competitive capitalism) the sub-capitalist modes may indeed be reinforced; in the second stage, however, capitalism 'takes root' and destroys all forms of commodity production outside of capitalism. Labour-power everywhere becomes a commodity. Rey (1973) in particular argues that for Marx and Lenin finance capital could take root in sub-capitalist modes and substitute itself for the production of subsistence goods. Hence, for them, the articulation was not a permanent feature, but only a transitory complement to capitalism. What is missing in their analysis, however, says Rey, is the crucial 'transition' achieved by colonialism, which represents 'finance capital's mission to create the conditions where the exchange of money for the living labour of workers can take place'. On the other hand Rey admits that sub-capitalist modes are not 'destroyed so thoroughly' as at the parallel stage in the metropoli; and this is because to expel the peasant from his land was just not possible, and because capitalism had occupied only the superstructure, not developing agricultural production.

Paraphrasing Rey is certainly not the fairest way to present his case, but he is wrong, if only because he supposes one unique form of capitalist development, and does not admit of a genealogy of capitalism. More serious, he seems to believe in 'parallel stages' of development (an idea that was discredited with Rostow) and does not develop an objective analysis of the mode of reproduction of the social division of labour in what is here called authoritarian capitalism. He simply assumes that the expropriation of the peasantry will take place. Finally, of course, it may, but the transition may be delayed for centuries. As Bradby says, 'essentially "development" is about widening capitalist relations of production and increasing the labour-base from which surplus-value can be extracted. In today's "underdeveloped" countries capitalism is still at the stage where it can only widen the labour-base by reinforcing pre-capitalist relations of production' (Bradby 1975). What Rey cannot explain is how, in fully fledged monopoly capitalism, the majority of the world's population, whether in colonies or neo-colonies, are still producing as peasants. The peasantry as a whole has not been destroyed, has not 'disappeared', but continues to be confined within a certain social space which is the sub-capitalist modes of production.

The appropriation of the peasant surplus

Modes of production are articulated so that value may be transferred from sub-capitalist modes to the dominant capitalist mode. This value represents the surplus-product of the peasantry, and before it can be transferred this surplus must be appropriated. In large part the appropriation takes place through different rent mechanisms, and this rent is common to all sub-capitalist modes (as defined here). Marx in the passage on 'pre-capitalist forms' of rent (Marx 1974) notes that this rent always derives from the same conditions, which are landed property in the possession of a ruling class who hold the direct producers in a relation of political subordination; direct producers in 'effective' possession of the means of production (as they must be in non-capitalist agriculture); and, as a consequence, surplus-product appropriated on the basis of 'extra-economic coercion' through rent, in the forms of labour, kind and money. This rent in its different forms (money = leasehold; kind = sharecropping; labour = labour-service obligations) is clearly a money-rent and equally clearly is a forced rent, which depends on 'extra-economic coercion' of peasantry on the land. Such a rent is a quite different category from capitalist ground-rent which in Marx's original analysis presupposes capitalist agriculture (and equally presupposes the 'disappearance' of the peasantry in the original form of capitalism). In this development there is a monopoly of land (one half of the 'double-mill'), capitalist competition to enter agriculture (given the low organic composition of capital in that sector) and rent is that portion of total surplus-value which accrues to the landlords, and which depends on the legal impediments to the free movement of capital created by the monopoly of land. As Lenin warns us (Lenin 1967)

> A strict distinction must be drawn between money-rent and capitalist ground-rent: the latter presupposes the existence in agriculture of capitalist and wage-worker; the former the existence of dependent peasants. Capitalist rent is that part of the surplus-value which remains after the deduction of the employer's profits, whereas money-rent is the price of the entire surplus product paid by the peasant to the landowners.

In the historical development of authoritarian capitalism the peasant surplus is of course extracted through money-rent. There is monopoly of land, but in general no capitalist competition for land and this for two reasons. In the first place the potential rate of profit in the countryside is low, and certainly lower than that existing in industry or real estate. Despite the low organic composition of capital in the countryside, capitalist exploitation of labour is not profitable precisely because it has to compete with the super-exploitation of labour in the sub-capitalist modes – and wages cannot be paid below subsistence. Exceptions to these constraints can emerge where an alteration in the production function reaps huge rewards in increased productivity. In the second place, the landlord and 'capitalist' rentier are often one and the same person. As Marx made clear, this landlord may play according to capitalist 'rules', and make every effort to maximise profit, but he is not in fact a capitalist. 'The historic process is not the result of capital, but its prerequisite. By means of this process the capitalist then inserts himself as a (historical) middle man between landed property, or between any kind of property, and labour . . . but . . . if we now talk of plantation owners in America as capitalists, if they *are* capitalists, this is due to the fact that they exist as anomalies within a world market based upon free labour' (Marx 1964). In other words, while the rent received by landlords may be distributed by capitalist relations of distribution, it is generated within a labour-process dominated by non-capitalist relations of production.

The different arguments against this position are all fallacious in one way or another. For instance it is argued that if land has a *price*, which is understood as capitalised (capitalist) ground-rent, then this ground-rent must exist. But Marx theorised prices as capitalised ground-rent in the case of capitalist production in agriculture, and to assert that prices prove the existence of capitalist ground-rent is to beg the question. Here Rey is quite correct to condemn 'price of land' as an irrational category which has the function of subordinating the sub-capitalist modes to the capitalist mode, and behind the category always lies the reality of money-rent. In short 'price of land' is a function of the articulation. Then it is argued on a more empirical level that agrarian reforms are instigated by industrial bourgeoisies in the third world in order to reduce the relative participation of 'rent' in total surplus-value. The theoretical myopia of this assertion fails to perceive the emptiness of the 'evidence': not only are most agrarian reforms in the third world very limited, cosmetic exercises, but in most countries at most times a transfer of value from countryside to city can be clearly demonstrated.

None of this denies that conjunctures do occur in third world agriculture where better placed producers and/or producers on more fertile soils realise a greater than average profit. But this usually occurs in conditions of labour-scarcity which demand the introduction of wage-relations in the countryside; it is then perfectly correct to theorise this greater than average profit as a differential rent. The classic case where rent of this kind was important is Argentina, and Argentina is unique in Latin America for the historical absence of a peasantry. But if 'bigger than average' profits are realised in the sub-capitalist modes, then they are precisely that – a super-profit and not a differential rent. This follows from the logical premises of Marx's analysis. Differential rent is created, in the first instance, by a rising or fluctuating demand, such that less fertile or more distant soils are brought into cultivation (or more capital invested on some of the soils). Rising or fluctuating demand in the sub-capitalist mode however will induce the landlord as a 'maximiser of profits' to put more land into production *within* the unit of production, which is the large landed estate, such that the greater profits he realises represents an increase in money-rent, and not a differential rent. Historically it is true that any super-profits created in this way accrue more to commercial capital than to the landlord; what remains to the landlord will simply be looked upon as a 'bonus'.

Thus Wolf's original definition of peasants as rural cultivators who pay rent (Wolf 1966) seems substantially correct, if this rent is seen as a mechanism for siphoning surplus (for getting labour-power at a cost below its value). Not only peasants, but 'independent' petty commodity producers contribute to the main fund of rent if they are obliged to offer their labour to the landlord; where they are not, then their surplus-products may be extracted through market mechanisms of unequal exchange (that is directly to commercial capital) or through taxation (that is directly to the State). No doubt some independent peasants escape these different mechanisms of extraction, and succeed in capitalising their enterprise; such 'middle' peasants are inherently unstable, if unsubordinated, and always tend to merge into the main configuration of exploiters and exploited (compare Lenin 1967).

The dominance of the political

It is evident that the appropriation of the peasant surplus would not be possible without the political subordination of the peasantry, and the concept of a dominance of the political refers in the first instance to the specific role of the State in imposing this subordination and in achieving the general conditions for authoritarian capitalist development. But this role should not be conceived as an intervention of the State from outside the productive process, but rather as the presence of State in the form of institutions of labour control, which are at the same time the productive units of the sub-capitalist mode of production. In other words this State should be understood not only as a centralised and hierarchical apparatus, but as diffused throughout the countryside in large landed estates (in the Brazilian case the authoritarian capitalist State took several decades to reach any degree of centralisation). The different forms of estate (hacienda, fazenda, plantation – all institutions of labour control) are at the same time State apparatuses of political and ideological control, which are even seen as containing a specific 'culture of repression' (Freire 1972; Huizer 1973). In short, the dominance of the political is vested in the productive units of the sub-capitalist mode and through these institutions achieves the reproduction of the social relations of production of this mode (the reproduction of the conditions for the appropriation and transfer of the peasant surplus) and the reproduction of the peasantry as a class. But while it is clear where the subordination is imposed, it remains to be established how it is imposed: what is the exact place of the political in the process of reproduction?

Until now it has been suggested that the subordination is imposed by 'extra-economic coercion', but in Marxist theory such coercion is only extra-economic in the narrow sense of existing outside the sphere of commodity exchange; in other words it is a coercion unlike that exercised through the market in capitalism itself (Laclau 1975). Thus the concept of extra-economic coercion simply says that coercion exercised in non-capitalist modes is not capitalist, and, moreover, by projecting into the discussion of the non-capitalist mode a type of social rationality only existing in capitalism, it can confuse the conception of this non-capitalist coercion. For the non-capitalist coercion is clearly *economic* in the sense that it takes place at the level of production (as in the phrase 'the economic is determining in the last instance'), as well as being *political* (and so not only economic as in capitalism). But the concept of a political coercion can mislead as much as extra-economic coercion, for it only occasionally implies the use of direct coercion. If there were no monopoly of land, and the direct producers were producing independently, then they would certainly have to be coerced directly to give up part of their product. But, on the contrary, land is monopolised, and this monopoly creates the dependence of the peasants (which is usually sufficient to secure the peasants' surplus) and therefore the conditions of appropriation of the surplus. The amount of money-rent paid by a tenant, for instance, is usually agreed before the beginning of the agricultural year, and by definition such dependence is 'institutionalised' within the large landed estates. Thus the coercion is economic in that it is exercised at the level of production and the labour process, and political in that it is exercised through the monopoly of land, and the 'immobilisation' of labour. Rather than being 'extra-economic' the coercion is indissolubly political and economic.

In broad terms this conceptualisation of the coercion points to the dominance of the political in the process of reproduction, without denying that such political dominance is transmitted through economic structures which are 'determining in the last instance'. But it has been questioned whether there is any theoretical basis for such a conclusion. Hindess and Hirst (1975) in particular, in their commentary on Marx's observations on 'pre-capitalist' forms of rent (see above) assert that in Marx 'pre-capitalist rent is not in fact conceived as an economic relation at all, but as a relation of political domination'; that the intervention of the State is necessary to the extraction of rent, and that the State is a condition of its existence. In other words, there is no theory in Marx which establishes these pre-capitalist relations of production as relations of appropriation of surplus, which, in their turn, require the existence of forms of political domination as *mechanisms* of appropriation. Production therefore has no social conditions of existence, and the political and ideological instances, must be supposed to constitute the social relations of production. In the absence of any mechanisms of appropriation of surplus which arise out of the system of production itself, this is tantamount to an assertion by Marx of the dominance of subject by subject (and a 'conflation' of the level of the political instance with that of the political subject). Therefore, they conclude, the 'dominance of the political' is not executed over the economic instance, or through the structures of the economic instance: the dominance is one of 'political' subject over 'economic' subject.

In attempting to solve the difficulties created by their critique, Hindess and Hirst point to what they consider to be a key distinction, by which they contrast that *domination*, whether political, legal or ideological, which provides the pre-conditions of a mode of production's development, and that *dominance* which provides the conditions for the existence of real appropriation through the social relations of production, which in turn provide the social foundation for the political instance by creating the division between exploiter and exploited – a division internal to the system of production. But the distinction they make is not analytical or theoretical but historical (but as history is excluded from their book the distinction must enter as theory). It is moreover a distinction which would have emerged more clearly from a closer reading of Marx, and especially the key passage which asserts (Marx 1974) that 'in all forms in which the direct labourer *remains* the "possessor" of the means of production . . . the property relationship must simultaneously *appear* as a direct relationship of lordship and servitude' (my emphasis) – which is purported to demonstrate a 'conflation of political instance with political subject'.

The point to note is that the direct producer remains a 'possessor'. After what? After the expropriation of the direct producers – which is in plain contrast to what occurs during the stage of primitive accumulation in the emergence of the original form of capitalism. Expropriation may not only indicate this direct divorce of petty commodity producers from the land and their impulsion to the cities, but also indicate, as in the sub-capitalist mode of production, the prevention of a direct and un-mediated combination of labour with land. With the direct producer remaining in 'possession' the property relationship appears, but only appears, as a direct relationship of lordship and servitude. In fact, the monopoly of land already achieved provides the conditions for the appropriation of the surplus through the relations of production – the 'social foundation of the division of exploiters and exploited'. And this monopoly (private property in land) is not merely a juridical relationship (which supports the social relations of appropriation of surplus as in the capitalist mode), but is also the specific units of production and at the same time the institutions of political control of the sub-capitalist mode of production. (Hindess and Hirst later make a similar point by emphasising that the landlord does not merely have legal rights, but is also engaged in the process of production, controlling allocation of the means of production and the division of labour.) In short, there is no necessary theoretical 'conflation', or certainly none in Marx, of economic relations with the political conditions of their existence.

Modes of production and the pioneering peasantry

This presentation and analysis of the general conditions of accumulation in the Brazilian countryside facilitates a more accurate appreciation of the role of the frontier in the formation of the national economy. It is now clear that the primitive accumulation practised on the frontier is just one part of a more general process of accumulation in the countryside, which proceeds in subordination to the dominant capitalist mode of production, and that the transfer of value from the frontiers is additional to the more general transfer from sub-capitalist modes of production. At the same time it is the process of frontier expansion which extends the scope of accumulation in the countryside, and establishes this 'sub-capitalist environment'. In this connection, it is clear that the articulation of modes of production in Brazil did not mean the transformation or subordination of already existing pre-capitalist modes of production, but the dynamic *creation* of a 'sub-capitalist environment', first in response to the pressures of commercial capital, and later to the penetration of capitalist social relations of production. In this century the creative force has been the frontier. The continual reversion on the frontier to the dominant duality of latifundio–minifundio, determined by its own conditions of development, has itself extended and reproduced the essential condition for accumulation in the countryside which is the monopoly of land. In this way the 'freedom of the frontier' leaves as a legacy further institutions of labour control, and the mass mobility of the pioneering peasants recreates the conditions for the mass immobilisation of labour in such institutions. In short, the frontier expansion does not merely contribute to the accumulation in the countryside itself, it reproduces and extends the conditions for it.

Thus it is evident that the pioneering peasantry contributes massively, directly and indirectly to accumulation in the countryside, but the peasants themselves produce within the petty commodity mode of production which is subordinate both to the dominant capitalist mode and to the sub-capitalist modes. The frontier peasant may be exploited by commercial capital, the vassal of capitalist industry (even to the point where not only his surplus-labour but also his necessary-labour time is appropriated) and so contribute to lower the cost of reproduction of the labour force in the capitalist mode, allowing labour-power there to be bought at a cost below its value; similarly in working on a neighbouring estate or capitalist enterprise, with part of his subsistence guaranteed by his frontier farm, or clearing land on the frontier in return for the privilege of planting subsistence crops for a year or two, the frontier peasant offers his own labour at a cost below its value. And given his precarious social position, which places him outside both capitalist wage relations and the paternalist relations of the large estates, this value is extremely debased.

Apart from all this, while the pioneer peasant creates the conditions for others to 'possess' the land – even if the 'possession' is predicated upon exploitation through sub-capitalist relations of production – he himself is repeatedly dispossessed of his land, and driven off it. And the extremities of violence he suffers in the process are not equivalent, in theory or practice, to the economic–political coercion exercised within the established sub-capitalist environment. This latter coercion is exercised in order to preserve the general conditions of accumulation; the violence to impose those conditions against the will and interests of the petty commodity producers on the frontier. In the same way, if the 'political' is in some sense dominant in the sub-capitalist modes but operates through the economic structures (the units of production) in order to achieve the reproduction of the relations of accumulation (and if the coercion is an 'expression' of that dominance), the intervention of the 'political' on the frontier knows no such transmission, and the pioneer peasant suffers the naked force of the State, or the State's sufferance of private violence.

The contemporary context

In recent years, with the massive growth in the influence of monopoly capital corporations, developments have taken place which might be seen to call in question the relevance of this analysis. The patterns of investment in the periphery, for example, have shifted rapidly into the production of strategic materials like oil, and into manufacturing production, and overall allocation of investment funds has shifted from periphery to centre. At the same time selective areas of agriculture in the periphery have been capitalised and wage-relations introduced. These latter developments are best summed up in what has come to be called the Green Revolution. This 'revolution' has been promoted from the centre speciously to bring high productivity farming to the periphery and so solve the world's food shortage, and 'peripheral' States accept the initiatives because investments in 'select areas' help them guarantee the steady supply and high quality of industrial raw materials. They themselves promote high productivity production of exchange earning crops (or crops which are costing a disproportionate amount of foreign exchange). In fact, as far as the centre is concerned, the 'revolution' is more directly designed to create a market for the agricultural chemicals and hardware produced by the monopoly capital corporations (Feder 1971), and where 'peripheral' States are unwilling partners they are convinced by 'aid' incentives. But, whatever the motives, the 'revolution' is transforming regions of sub-capitalist production into areas of capitalist agriculture. Is this the beginning of the final 'dissolution' of the sub-capitalist environment?

Dissolution of sub-capitalist modes has occurred repeatedly in the past, but this dissolution has never been more than partial. Even today, in the great majority of cases, capitalist penetration of this kind is very limited, in terms of geography and space, and highly selective, in terms of produce, soils and technology. This is still true for instance of India, one of the 'targets' of the Green Revolution. As Alavi writes, 'if we look at the developments in terms of rates of growth, the change appears to be quite dramatic . . . but the absolute magnitude of the changes must be considered against the background of the enormous size of the Indian countryside . . . there is still a very great deal of ground to be covered before we can see it as having made a qualitative change' (Alavi 1975), and the same might be said of the penetration of capital into the Brazilian countryside, and more especially into the Amazon Basin. But this is not to deny that, following the transition to the dominance of monopoly capital in the countries of the periphery, the relative 'weights' of different modes of production in the economic system may change, as may the relative importance of the articulation to the expanded reproduction of the system itself. These changes may mean that the amount of surplus transferred from the sub-capitalist modes is relatively smaller (in comparison with the overall rates of accumulation in the economy) but, equally, and not necessarily in contradiction to this, the modifications may be understood as maintaining or increasing the *rate* of transfer from countryside to city – and the transfer may be just as essential as previously to the reproduction of social relations of production in the economic system as a whole (and also essential to the high profits generated by some multi-national companies). Thus, these changes should be theorised not as a dissolution of the sub-capitalist environment (for the partial dissolution in some areas may imply a further extension or reinforcement in others), but rather as a secular change in the *form* of the articulation, in answer to the changing needs of capital accumulation in the social formation as a whole.

Thus the 'co-existence' of the different modes of production continues though the form of articulation changes. It has been seen, principally, in the Amazon, how capitalist enterprise is pushing back the frontier, which now carries not only a primitive accumulation, but the seeds of capitalist wage-relations and the creation of surplus-value. To this end, State agencies are not only providing incentives to capital, but, by seemingly contradictory policies, are creating potential 'labour reserves' in the region. But equally it has been seen how it is still petty commodity producers, by their activity on the ground, who clear most of the land; and how even large capitalist enterprise itself creates and extends the sub-capitalist environment by the imposition of sub-capitalist relations of production. The new form of articulation is here seen encapsulated within the enterprise. Capital pushes to the furthest reach of the frontier, but instead of capitalist 'rationality' promotes slavery, servitude, and, still, the violence of primitive accumulation.

Given these contradictions in the process it is not possible to predict the social results of the changing form of articulation. But it may be that where capitalisation of the countryside is relatively widespread and the traditional institutions of labour control (the social and economic bases of the political subordination of the peasantry) threatened with 'dissolution', then, in the right political conjuncture, the peasantry may be far enough mobilised themselves to threaten the bases of labour control in the countryside. This may contribute to impel a radical change to a far more highly centralised authoritarian State, where control and coercion of labour in the countryside and city is exercised by the central State apparatuses. If this picture has any plausibility, then in Brazil it was presented early in the 1960s, with the mobilisation of Peasant Leagues throughout the countryside, and especially in the North-East. To some degree this may have precipitated the installation of the highly centralised military–authoritarian State – and it is this State which has ordered the occupation of the Amazon. This is material for the next chapter.

10

The frontier and the formation of the Brazilian State

Historically a continuing political intervention has been required to reproduce the conditions for accumulation in the Brazilian countryside. In particular it is the different institutions of labour control in the countryside which have effectively contained these conditions through the monopoly of land and the subordination of the slaves or peasants. This pattern of extensive monopoly of land was established during the colonial period. In contrast to Spanish America, the late discovery of precious metals led to a conquest by 'colonisation', and sugar production in *engenho*, not silver mines, shaped the settlement of Brazil (Halperín 1969). At the same time the Portuguese colonial administration commanded relatively fewer resources than the Spanish, and the consequent devolution of power to local landowners led to a socio-juridical division of the land which was far in advance of economic demands (Prado 1962a). However, these demands increased dramatically after the transition from production for a colonially controlled market to production for a world market which by early in the nineteenth century represented a generalisation of appropriation of surplus through exchange relations dominated by commercial capital. At the political level this transition was from control by colonial administration to control by an autonomous Brazilian State: given the political conditions for accumulation, this control was inevitably vested in the State apparatuses in the countryside, which made this incipient State a landowners' State.

During the Empire the landowners were both the dominant class and the class-in-charge of the State (Poulantzas 1968). Despite the formal political organisation of the Empire, the central State at this time achieved little projection. Regional revolts in the 1830s and 1840s proved extremely difficult to put down, and throughout the life of the Empire the 'provinces' retained considerable autonomy. Following the fall of the Empire it was still true that the landowners ran the local states, and the most powerful states ran the Republic: by the new Constitution of 1891 all unclaimed and untitled lands passed under the jurisdiction of the local states, and this control of unclaimed land was of course important in a country where the motor of economic growth was the expansion of the agricultural frontier –

cattle and especially coffee (Furtado 1963). But the growth of the coffee sector and the progressive integration into the capitalist world market also required, and created the conditions for, a greater centralisation. This centralisation – what Cardoso and Faletto call the increasing dominance of the city over the countryside (1969) – had begun under the aegis of the Imperial State, and resulted not only from economic but also from political developments, such as the militarisation required to wage the Paraguayan War. But, above all, the process was accelerated by the expansion of the coffee frontier in São Paulo, and the consequent introduction of capitalist production relations in the countryside (Martins 1975). The signatory to the international financial agreement at Taubaté in 1907 was the local state of São Paulo, and this is indicative of this acceleration.

The incipient centralisation of State provoked severe political contradictions. Brazil had suffered no political rupture at the time of independence (Halperín 1969), and so had not immediately imported the liberal institutional forms of State so favoured by many Spanish American neighbours. However, laissez-faire ideology had free currency and seemed to correspond to the economic realities of the dominant class: a theory of comparative costs evidently appealed to a class whose wealth derived from the export of sugar, gold, rubber and especially coffee. So there was a clear logic in the imposition of liberal, 'representative' institutional forms in the First Republic. This 'representation' was very restricted but even because of this served at least temporarily to reconcile divergent regional and sectoral interests. The local states' autonomy was preserved by the *política dos governadores* whereby the different local state governors supported and 'legitimised' the President on the understanding that there would be no Federal interference in the local states (Pereira de Queiroz 1969). In practice this compromise favoured the most economically powerful states of São Paulo and Minas Gerais (*política dos governadores do café com leite*), and by an 'informal agreement' it was their governors who alternated in the Presidency. In this way the elaborate compromise overcame the contradictions in the short term, but with executive prerogatives debated between local states and Federal State they would inevitably reappear.

This inevitability derives from the emergence of an agrarian, and later an industrial, bourgeoisie in São Paulo (Dean 1969). In other words the contradictions within these speciously liberal institutions of the State became yet more acute with the emergence of capitalist social relations of production as dominant in the economy. It is true that similar institutions within the liberal State in the central capitalist countries had effectively represented the 'divergent' interests of the 'many capitals' of the period of competitive capitalism (Mandel 1976); but the role of the State in Brazil at the beginning of the twentieth century was very different. Given the special place of the 'political' in promoting the expanded reproduction of the economic system, the State had now to respond to the new production structures and ensure their dominance in the social formation. The consequent range of political supports to entrepreneurial activity is extensively documented (Dean 1969; Baer 1965; Bergsman 1970; Simonsen 1973; Furtado 1963).

But the 'liberal' institutions of the State proved inadequate to the task: not only did they not reflect the authoritarian social and economic bases of the State, but they lacked the cohesion, and, especially, the concentration of power in the executive to implement the measures which would guarantee this dominance. The situation is something like an attempted intervention on the scale of the Absolutist State in Europe, only with the political institutions of the much later liberal State. As a consequence of the inadequacies, the contradictions provoked by a necessary centralisation escaped the confines of the 'liberal' compromise. Given the concentration of capitalist production in São Paulo it was inevitable that the crisis of centralisation would take on a *regional* character: in the face of a deepening economic crisis in the coffee sector this state broke the agreements and bid for sole control of the Federal executive. The general crisis of State which follows found expression in the Revolution of 1930. Whatever else this Revolution signified it was certainly a bourgeois revolution, in the sense that the new form of State which emerged in the 1930s was clearly a capitalist form of State.

Forms of the capitalist State

The general function of the State in class societies is to maintain the unity of the social totality, and actively prevent the process of social disintegration promoted by class struggle (Hindess and Hirst 1975). As classes themselves are defined by their integration into the internal relations of production of the society, which achieve the appropriation of producers' surplus by non-producers, it follows that determinate types of State correspond to determinate production processes in different societies. In other words, for every mode of production there will be a different type of State, and the role of the State in the reproduction of the productive base of society is different in different modes of production. Capitalism as a mode of production is characterised by the separation of the direct producers from the means of production, and the generalisation of commodity exchange including, crucially, labour-power as a commodity. The labourer must sell his labour-power in a market, and the impersonal character of the exchange is the essential mediation of the process of exploitation in capitalism. Just as the labourer is 'free' to sell his labour-power, so the capitalist is 'free' to buy it, and 'free' to compete in the market for the surplus-value which only 'free' labourers can produce. Given the presence of these 'individuated individuals' (Marx 1975), capital can rule impersonally through market competition, instead of imposing personal relations of dependence and domination as in most pre-capitalist modes of production. In these conditions the capitalist State appears autonomous vis-à-vis the process of production: this relative autonomy is an objective characteristic of the capitalist State (Poulantzas 1968), but insofar as it appears 'independent' and 'above society' the autonomy also fulfils the ideological function of concealing the class nature of the State. What this autonomy means is that there is no permanent and direct intervention in the production process, but the State still provides the conditions for the reproduction of the relations of exploitation.

It does this through its fundamental role of mediator between the 'bare individuals' and the totality which is the society; between individual competition and the process of production which is increasingly socialised. The State recognises, socially and legally, the individuals as separate, free and equal members of a society, who together, at the same time, compose the totality which is the 'nation'. So each individual is given the political rights and duties of the 'citizen', and the citizens compose the nation. In this way, antagonistic classes are transformed at the level of the State into a non-antagonistic totality, made up of the sum of individual citizens who hold 'power' through the mechanisms of universal suffrage and citizens' rights (Poulantzas 1968). These are the institutions of the classical liberal form of the capitalist State; but in general the various forms of political participation in capitalist society have as their basic requirement this process of 'individualisation' at the level of production, and the further 'individualisation' through the mediation of the State. Despite the mediation, the State still has a class character, which is to reproduce the bourgeoisie as the dominant class. This bourgeoisie by definition is a class which competes 'internally', and requires in the State a centre of direction which will unify the class but at the same time preserve its essentially competitive character (Mandel 1976). Therefore insofar as the State is an 'executive committee of the bourgeoisie' it tends to assume a range of elective, representative and deliberative institutions. As the bourgeoisie itself is fully engaged in making profit (the greatly increased social division of labour in capitalism makes bourgeois economic activity very complex) the executive functions of State are entrusted to specialised social categories like the bureaucracy, which further creates the appearance of the State as separate from the bourgeoisie. The hegemony achieved by the bourgeoisie in the original form of capitalism depends precisely on this mediation of the State, which allows the bourgeoisie to 'present itself as an organism in continuous movement, capable of absorbing the entire society' (Gramsci 1973).

This same type of State will exist wherever the capitalist mode of production is dominant, but the form of State will change where the overall process of accumulation is changed. The process of accumulation in peripheral formations such as Brazil is necessarily different, as was apparent in the investigation of its internal relations of production. Here the growth of the capitalist mode of production has as the condition of its own reproduction the growth and extension of a sub-capitalist environment. Far from accomplishing the generalised 'individualisation' of 'free' labourers this capitalism supposes the reproduction of institutions of labour control which directly subordinate the producers through 'political–economic' control (compare Chapter 9). In these conditions there exist clear restrictions to the classical mediation of the capitalist State (which is largely confined to the urban context) and, in fact, the State fulfils the contrary functions of labour control and extraction of surplus-labour through direct and permanent intervention in the productive process. In other words as the reproduction of the economic system is not determined solely by the laws of capitalist accumulation, so this form of State cannot logically resemble the classic liberal form. Its role, it has already been suggested, is not only to guarantee the reproduction of the conditions of a capitalist accumulation proper, but to reproduce the articulation of different modes of production, and indeed guarantee a continuing primitive accumulation on the frontier, which itself reproduces the articulation. Therefore the capitalist State in Brazil does not even *appear* as autonomous and separate from the productive process, and, in controlling labour in the countryside (to talk only of the countryside here) and excluding that labour from political participation, obviously denies to the peasant the status of citizen. Peasants do not have 'equal' rights and duties, but are politically subordinated (most often in institutions of labour control) and ideologically isolated in that they are subject to a different political 'discourse' than that offered to the citizens. Thus the bourgeoisie may become economically dominant, but can never be hegemonic: not only because of the 'dominance of the political' in the reproduction of the composite internal relations of production, but because its rule cannot be mediated in classic form to the 'subject-individuals' of the society – although the adoption or importation of liberal institutions of whatever kind may represent bourgeois attempts to gain hegemony despite the structural constraints. Such attempts are doomed to failure.

The overall role of the State in the peripheral capitalist formation includes specific particularities already raised in the discussion. For instance, the private violence exercised on the frontier was interpreted not as a failure of the State to achieve a monopoly of the means of coercion in the national territory, but as a reflection of its paramount political task of underpinning the appropriation and transfer of value across different modes of production: private violence is not repressed as long as it contributes to the extension of the sub-capitalist environment and the reproduction of the articulation of the different modes. This articulation of modes within the economic system exists because of the reflexive nature of capitalist growth in the periphery (which is a response to an autonomous process of capitalist expansion elsewhere) and the 'penetration' of capitalist social relations of production into the periphery from the centre.

Within this world system the peripheral economic system is 'dependent' and, similarly, the overall role of the State in the peripheral capitalist formation can be assimilated to the idea of a 'dependent' State. As capitalism becomes entrenched as dominant in the peripheral economy, so that economy becomes more closely dependent on the world system: the role of the State in the appropriation, concentration and transfer of value implies its 'dependent' position insofar as the transfer of value now supposes a final transfer to the capitalist centres – through mechanisms such as unequal exchange, repatriation of profits, royalties and insurance, debt servicing and amortisation, illegal capital flight etc. (Griffin 1969). The further the State becomes 'dependent' the further its role in guaranteeing the articulation (and so guaranteeing higher average returns on foreign capital investment and higher rates of profit for multinational corporations) is reinforced; and the further that role is reinforced, the more 'dependent' it becomes. Given this principal role, and its reinforcement, the ways in which economic exploitation is mediated through the State to the direct producers is very different from those in a liberal capitalist State. Violence is one of the mediators, and, later, others will be examined as characteristic of the *authoritarian* capitalist State.

The State-in-formation

While it is important to distinguish these different forms of capitalist

State, to leave the discussion there would be to lose the special perspective on the State that a study of the frontier can create. It may be that the 'dominance of the political' in the authoritarian capitalist formation correctly implies that the 'political State' develops in advance of a differentiated 'civil society', but it should not imply, equally, that this State is fully present and fully centralised from the first moment of its historical existence. Moreover the problem is general in that current theories of the State tend to present their logical and 'structuralist' interpretations as if they corresponded to a constant and unchanging reality. Of the major theorists Poulantzas (1968; 1974b) comes first to mind in this connection, although no one would wish to deny the importance of his contribution to the theory of the capitalist State; Anderson (1974) on the other hand, in his excellent study of the Absolutist State, succeeds in escaping this tendency almost completely. In general, however, as Blok perceives it (1974), 'even those scholars whose investigations are guided by a long-term historical perspective are primarily concerned with formal definitions, attributes and reified abstractions. They often speak of State as if it emerged full-blown with all its attributes and armour.' The alternative to this is to present a dynamic conceptualisation of the secular formation of the State in an attempt to see how it has become what it is and what it is becoming. This would involve different 'ideas' of the State (State-in-formation; bureaucratic State; political State) which should be understood as progressive 'approximations' to the concrete reality, and not as mutually exclusive theoretical categories. Different 'spill-overs' are likely in the discussion, as the 'approximations' all relate to the relationship between the State and the frontier.

It has been argued that the role of the State can only be analysed in terms of the expansion of the national economic territory through the monopoly of land and the extension of a sub-capitalist environment. As an economic process this can be correctly viewed as a pattern of horizontal growth, and 'lateral' accumulation, through the appropriation of value across articulated modes of production. In less abstract language, more and more land is taken into production by an extensive and largely non-capitalist agriculture. But insofar as this expansion implies the control of peasantry through 'political–economic coercion' within State apparatuses it is also a political process, which can be conceived paradigmatically and geographically as an expansion of the 'political centre'. This political expansion takes place in the first instance through the units of production which are simultaneously the social and economic bases of the authoritarian State, and which constitute the initial extent of the political intervention in the productive process. This means quite simply that economic expansion and political integration of the national territory go hand in hand. These political and economic processes compose the secular formation of the State in Brazil. It is therefore correct to theorise the formation of this State, like any other, as based on the internal relations of accumulation.

The secular formation of the State as an expansion of the 'political centre' has led to the early stage of State formation in Brazil being characterised as a 'diffuse authoritarianism' (Velho 1976), which evidently corresponds to more general interpretations of *coronelismo* (Nunez Leal 1975; Pereira de Queiroz 1969), and to classic theories of the 'patrimonial State' (Weber 1957, 1970). As late as the First Republic, as observed above, the polity was characterised by a marked autonomy both of the local states and of the local landowners themselves. At the same time it was suggested that the State became slowly but progressively more centralised and this was seen as a response to the growing need for regulation and organisation of a fast-growing economy; in other words as a response to the emergence of capitalism as dominant in the economy (Meireles 1974).

The most rapid progress in this direction was made between the fall of the First Republic and the establishment of Vargas' Estado Novo. It is almost a commonplace that following the fracturing of the export economy in the years after the world crisis of 1929, the loss of hegemony by the landowners was not compensated for by the emergence of any new hegemonic class or class fraction (Ianni 1964). More than one group in the society attained an 'economic–corporate' identity, or the ability to defend its own interests, but no one group achieved the political and civil ascendency to make it capable of imposing its own economic project and re-integrating the society around it. Thus the authoritarian capitalist State of Vargas did not mediate the hegemony of a new *clase dirigente*, in Gramscian taxonomy, but was rather founded on a semi-corporate alliance of economic sectors forged in the mould of a traditional paternalism, and articulated through the State itself. It is the State which now achieves more 'autonomy' of regional and local interests through its role as arbiter of class and regional interests. To a greater degree than ever before it can now impose certain priorities and policies 'throughout' the national territory.

The importance of the shift which occurred following the Revolution of 1930 is reflected in the dominant themes of Brazilian historiography since the 1930s, those of centralisation, bureaucratisation and militarisation. Other themes of urbanisation and industrialisation reflect the growing dominance of industrial capitalist production, and consequently the dominance of the city over the countryside. But this does not mean that the countryside, and the institutions of the 'diffuse' State became suddenly unimportant. Vargas, throughout the 1930s, had to take landed interests (sugar, coffee) into account in his political calculations quite as much as the interests of the emergent industrial sector. Even after the failure of the 1932 'counter-Revolution' he continued to buy and to burn coffee to the value of 10% of the GDP, and one of the first Federal agencies created was the Institute of Alcohol (Furtado 1963). This political resilience of local interests is demonstrated by his reliance on the extensive use of 'interventors' who had to be regularly rotated as they became rapidly captured by such interests. Indeed the viability of the 'hegemonising' alliance depended precisely on a complex of compacts and alliances at regional and national levels, and much of the political history of Brazil following the fall of Vargas can be seen as a continuing attempt to maintain this alliance in the face of rapid economic change, and the mobilisation of a political public in the cities (Weffort 1965; 1966).

In terms of the two major movements which constitute the formation of the State – the expansion of the 'political centre' on the one hand, and the centralisation of the capitalist State in tandem with the transition to capitalist production on the other – it should be clear that these are not contradictory movements with the contradiction in some way resolved by the 1930s, but complementary movements, which became increasingly so after 1930. In short the period of most rapid centralisation and bureaucratisation is also the period of the most rapid expansion of the pioneer frontiers. Evidently there are many other major developments in play, some of which have been mentioned; but it is still correct to say that the State becomes more centralised in response, at least to a degree, to its role in guaranteeing the articulation of different productive systems in the countryside. The authoritarian capitalist State and its new centralised apparatus function to achieve the conditions for the expanded reproduction of the system as a whole and this includes, importantly, a continuous transfer of value from countryside to city.

The two movements are complementary in that one implies, indeed demands, the other, but not in the sense that the formation of the State follows a simple trajectory from the political centre, both horizontally and vertically. National systems of power can grow without obliterating local power holders, and in no sense is the expansion of the political centre to be conceived of as a unilateral projection of unifying and civilising central forces. In the first place the expansion is, by definition, a progressive monopolisation of the land, and this does not go uncontested. It has been argued in previous chapters that throughout the course of this expansion the peasantry is engaged, massively if unequally, in the struggle for land. In the second place, local and regional landed interests (for reasons of political autonomy and economic autarchy, or for the purposes of 'negotiation') (Mercadante 1965) can and do block the expansion of the centre, for greater or lesser periods of time. To some extent the central State apparatuses are extended in breadth and depth in order to overcome this order of resistance, and increase the range of State intervention on the frontier (this was documented in the investigation of the operation of the Federal agencies in Chapter 7). Insofar as State intervention and the growth of the State apparatus is a response to autonomous power locations on the periphery, which in their turn generate new private initiatives by the landowners and entrepreneurs on the frontier, then the process is dialectical in its development through time.

The idea of local resistance to the State by those who 'participate' in the State should cause little surprise. As Whitehead points out (1975), 'many, if not most contemporary States, contain feudos of semi-autonomous power, the existence of which may be passively accepted, but equally may be strenuously contested by the State apparatus'. In the case of Brazil, Furtado (1965) was convinced that 'since the regional political institutions are mostly strictly controlled by the old ruling class, the ability of the central authority to pursue its chosen lines of policy is consistently sabotaged by the obstructive action of local interests. To overcome these obstacles is a costly and difficult business.' In this essay Furtado focused attention on the period following the re-democratisation of the country in 1945, and demonstrated how the new parties betrayed all the common traits of the old *coronelismo*. Regional parties often wielded inordinate influence over policy at a national level, and at the least remained largely autonomous. In the same way the Federal administrative machinery never entirely escaped control by regional political groups, and, as always, the effectiveness of the central State often depended on its ability to 'negotiate' support in the localities and regions. During the period of his discussion the conflicts implicit in the 'political intransigence of the regional power centres when faced with the demands of the centralised State' found their highest expression in the deep political divergences between the President (voted in by the urban electorate) and the Congress (still captured by the landed interests, who controlled the votes of the subordinated peasantry). Once again, with the increasingly rapid construction of a 'parallel' bureaucracy to overcome the impasse this represented for policy-making and implementation, the resistance of local power holders is dialectically related to the growth and centralisation of the State apparatus.

Thus the movement to a more centralised State apparatus complements the expansion of the political centre in the degree that it grows through the dialectic implicit in this expansion. As an economic phenomenon the expansion can be viewed as a process of accumulation which generates its own contradictions, which an increasingly centralised State attempts to resolve (through its ministries, or autonomous and semi-autonomous agencies); as a political phenomenon, the expansion represents an extension and increase in political intervention in order to guarantee the now more complex articulation of modes of production. This guarantee requires different mediators from those of the classic liberal form of capitalist State, in order to constitute the peasantry as objects of exploitation in sub-capitalist modes of production, and in order to 'negotiate' and accommodate the economic interests of the periphery (periphery being defined as all local and regional interests related to the ownership and control of land). As already suggested, violence against the peasantry is one such mediator, but while violence serves for the politically subordinated, other mediators must constitute the landowners and entrepreneurs as 'citizens' who are 'free' to compete for profits and 'obliged' to participate politically in the nation-state.

The bureaucratic State

The increasing centralisation of the State-in-formation is manifested socially and politically in the growth of bureaucracy, which embodies the growing social organisation and political order of the State. Theoretically it has been proposed that the 'political State' develops in advance of 'civil society', and it is empirically proven that since the 1930s the Federal State in Brazil has created and consolidated what appears to be a prematurely large public bureaucracy. This is so far true that Brazilian politics, it is said, is permeated by a 'bureaucratic ethos' (Schmitter 1971). But the growth of bureaucracy has not necessarily led to a greater degree of control over the periphery by a more rational and cohesive political centre. Much to the contrary, despite the concentration of financial and material resources in the central State apparatus, the bureaucracy itself appears to escape unified control, and continues to take 'irrational' decisions and make 'irrational' responses, which reveal patterns of patronage, particular interests and political opportunism. This 'irrationality' of the bureaucracy is not so much the result of its size, as of a haphazard pattern of growth by accretion, without revision or reorganisation of its internal structure, which leaves it administratively uncoordinated and politically inconsistent. Yet however far the internal structures are investigated, bureaucratic malfunction and ambiguous operation are not finally explicable in terms of an organisational malaise, but only in terms of the political functions which the bureaucracy fulfils.

The argument of the previous chapters has demonstrated how the bureaucracy works to solve one of the principal problems of the State-in-formation, which is the cooptation of autonomous interest groups and capitals on the periphery, and in particular on the frontier. The expanding political centre needs to assimilate these groups and win their consent to the existence and expansion of the State. This it does by corruption, which is a constant in the power equation, but which is accentuated after 1964 precisely in the absence of any 'political process' (and when politics itself is viewed as corrupt). Bureaucracy is the carrier of corruption, not usually in the form of direct bribery, but in its role of promoting economic speculation and guiding economic 'entrepreneurship' – both of them focusing for the most part on speculation in land on the frontier, and on depredation of the unimproved resource base. The practice of this 'bureaucratic entrepreneurship' serves the clear political purpose of absorbing and accommodating the autonomous groups on the periphery. At the same time it inevitably diminishes the public 'effectiveness' of bureaucracy, as private interests infiltrate key offices and capture command posts within the uncoordinated apparatuses (and this is a result as well of the dialectical growth of State within the process of political expansion). But the 'ineffectiveness' and 'irrationality' of bureaucracy only appear important if the stated political objectives of the central State are given analytic precedence over the unstated political process of State formation on the periphery.

The effects of this 'bureaucratic entrepreneurship' are not merely political but economic, insofar as this kind of corruption concentrates resources, and especially land, under monopoly control of the interest groups and capitals on the periphery. It has already been observed how far bureaucratic 'inefficacy' masks an intense political activity which contributes to the continuing divorce of the direct producers from their means of production and the horizontal growth which is the process of primitive accumulation on the frontier. In general bureaucratic entrepreneurship is instrumental in achieving a certain rate of exploitation of labour on the frontier, and in achieving the extended reproduction of the sub-capitalist environment. In other words, the bureaucratic 'inertia' does not only perpetuate the general conditions for primitive accumulation, but contributes to achieve the appropriation of value within the sub-capitalist modes and its transfer to the dominant capitalist mode. In this broad conceptual context the damage that is done to military aspirations of integrating the national territory or guaranteeing national security weighs little in comparison with the appropriation, concentration and funnelling of value to where it can flow to the financial and industrial centres like São Paulo, or consume the products of monopoly capital.

The bureaucracy is therefore characteristically uncoordinated, with an assortment of agencies and departments, some of which are partially controlled at different times by local and regional interests. In these conditions the corruption can breed conflict not only between these interests, but, it has been seen, between diverse agencies and apparatuses of the bureaucracy itself. These conflicts, however, are contained for the most part within the legal procedures of the hierarchy of the courts – another branch of the ramifying bureaucratic structure. The courts are continually processing a plethora of law suits (the great majority treating of disputes over land) and in doing so they absorb the conflicting pressures by providing multiple channels for bringing together different groups, public and private, into the same political arena. There the conflicts are caught within the complex litigation which slows the operations of the agencies, but binds bureaucracy to itself, and blunts the clash of private interests seeking profit. Once within the arena, public or private interest, bureaucracy or capital, can negotiate in a common language which is the language of law. The authoritarian capitalist State insists on regular judicial procedures, and on the historical persistence of legal forms, which betray the bias of the system towards legal negotiation, already seen as 'an adequate equivalent of a more collective, political expression of interest conflicts' (Linz 1964). In this way all conflict, actual or potential, is to be contained within the context of law where (in the absence of a political process after 1964 for instance) it will be given an 'administrative' solution. Thus legal claims and litigation orient the bureaucracy in its work of cooptation and conciliation and, in this sense, legalism complements corruption by providing a certain 'consensus', where corruption achieves an overall 'consent'.

While litigation achieves this necessary consensus between capitals and bureaucratic apparatus, it also comprises part of the bureaucratic contribution to primitive accumulation. The legal disputes and delays both allow the frontier peasants to move on to the land to clear it, and later facilitate their divorce from the land, and the imposition of the traditional pattern of property holding. While there is no intention of further rehearsing these arguments here, it is worth noting that *on the ground* the law complements, and can even provoke, the violence of the primitive accumulation, whereas *in the courts* the law acts as the language of a negotiation which may lead to compromise or capitulation between the litigants, but always avoids violence. In this way legal negotiation epitomises the general pattern of political development in Brazil, where political conflicts have always been 'accommodated' as far as possible (Schmitter 1971), but where, insofar as these conflicts are caused by economic contradictions, the non-antagonistic pattern has been achieved through increased exploitation of the dominated classes and in particular the peasantry. Moreover, the 'horizontal' growth of the economy through lateral accumulation has acted to prevent any antagonism between the traditional landowners for example and the agrarian and industrial bourgeoisies. The considerable transfer of value from countryside to city in the years after 1930 was possible not only because of the exploitation of the peasantry but more specifically because of the continuing primitive accumulation. The expanding frontier and the reproduction of the sub-capitalist environment has created, precisely, an expanding-sum situation where potential conflicts between entrepreneurial groups and capitals can be adjusted incrementally and accommodated, quite literally, at the margin.

This non-antagonistic pattern is not of course a natural phenomenon but is achieved by the State, which binds together the different 'economic–corporate' groups in the society, in the absence of a hegemonic bourgeoisie which could impose its own economic and social project. Such a bourgeoisie would have created the conflicts so noticeable by their absence. As it is, the non-antagonistic pattern is the result of the alliance of economic–corporate groups articulated through the State. The State practices permanent political intervention in the economy precisely to this end (Meireles 1974), and thus the alliance is something more than the 'populist alliance' seen in different Latin American countries at different periods since the 1930s. Specifically, this absence of antagonism, and the clear dependence of very different classes and class fractions (including the rural landowners, the industrialists, the commercial and financial intermediaries of the export sector) on State support and State funding, has led Caio Prado Jr (1966) to suggest in his analysis a distinctive Brazilian mode of development which he calls 'bureaucratic capitalism' – a privatised version of State capitalism. Moreover, this 'bureaucratic capitalist development syndrome' corresponds in no small measure to Jaguaribe's (1969) 'cartorial State', to Weffort's (1966) 'State of compromise', to Faoro's (1975) 'patrimonial State', and to Schmitter's (1971) 'system'. In the present analysis, insofar as horizontal growth and lateral accumulation contribute to the alliance by clamping together landowners and bourgeoisie and at the same time promoting the dominance of capital over the landowners, the 'bureaucratic capitalist State' represents but one aspect of the authoritarian capitalist State, which intervenes to achieve primitive accumulation and underpin the articulation of different modes of production. For at the political level this is tantamount to articulating the alliance of dominant classes and class fractions in city and countryside under the dominance of industrial capital. In other words the concept of 'bureaucratic capitalism' refers only to the superstructure of the economic system, while that of 'authoritarian capitalism' includes both economic and political determinations of the reproduction of this system, which make 'bureaucratic mediation' by the State imperative. Interconnections and alliances at the political level reflect, though not in any direct way, the articulated economic structure of the authoritarian capitalist social formation. Needless to add, the idea of a non-antagonistic reproduction of the articulation does not include the reproduction of the peasantry.

The political State

It is suggested that the process of centralisation of the State, marked by a massive increase in State intervention in the economy, and by an increasing concentration of power in the central executive, is not an answer to abstract demands of economic development, but a response to a real process of political struggle. The direction of the struggle is defined by the structural impossibility of a hegemonic bourgeoisie, and the consequent necessity for a permanent political intervention to achieve and maintain the conditions for the expanded reproduction of the relations of accumulation. In other words, the descriptive processes of centralisation, bureaucratisation and militarisation have to be theorised in terms of the specific political functions of the authoritarian capitalist State, and these functions are founded in the contradictions created by the penetration of capitalist social relations of production into the social formation.

These contradictions provoked their first major political crisis in 1930, when the State had to fill a political void left by the loss of hegemony of the 'export oligarchy'. In the absence of any *clase dirigente* which could provide political leadership, the State constructed an alliance which could guarantee the interests of the bourgeoisie as a whole, and in particular of the industrial bourgeoisie. As the crisis had been directly caused by contradictions within the dominant classes, between the traditional landowning class and the emergent bourgeoisie, and between sectoral and regional fractions of this bourgeoisie, it was sufficient at this moment for the State to reconcile the interests of these classes and class fractions, and act as political arbiter between them. The success of the alliance depended on the State's ability to obtain the consent of the dominated classes to this project, or, at a minimum, maintain careful control of their political practices and participation. The working class in the cities was clearly coopted through the corporativist framework of conciliatory labour laws, which were collated in the *Consolidação das Leis do Trabalho* in 1943 (Ianni 1971), and their struggle in some sense pre-empted by the political initiatives of State. The peasantry simply did not appear in the political arena, but remained totally immobilised within the institutions of labour control in the countryside.

Thus the State in the 1930s did not have to respond to the struggle of what Gramsci would call 'fundamental forces', or clearly antagonistic classes, but to the struggle of divergent fractions of the dominant classes. Its formation and centralisation were therefore defined by its functions of reconciling these divergent fractions, and of favouring the industrial bourgeoisie by a series of special policies and the promotion of the transfer of value from countryside to city (compare Chapter 3). These functions it fulfilled largely through the bureaucracy and the different forms of 'bureaucratic entrepreneurship'. Such State activity is reminiscent of the pseudo-private initiatives, and specifically bribery and corruption, which Gramsci sees as instrumental in extending and cementing the social base of the Italian ruling class in the nineteenth century by the absorption of opposition and allied groups in a process he calls *trasformismo*. This process is 'characteristic of certain situations, in which the exercise of the hegemonic function becomes difficult, while the use of force would involve too many dangers' (Gramsci 1973). But such a solution depended on the less than fundamental nature of the struggle, which was completely changed by the time of the second major political crisis of the century, in 1964.

In contrast to 1930 the struggle of the early 1960s was defined by the confrontation of fundamental forces, and the polarisation of the political arena between the political practices of opposing and antagonistic classes. In the cities, the recrudescence of open class struggle in a series of violent strikes and lock-outs clearly escaped the confines of the corporativist structure, which appeared no longer capable of containing the economic struggle within a legal and administrative framework. In the countryside, for the first time in Brazilian history, the peasantry were mobilised on a national scale, which, while there was never any real possibility of a 'revolutionary' alliance between peasantry and proletariat, evidently changed the dominant classes' perception of the struggle in the cities.

A national mobilisation of the peasantry has always had revolutionary implications in Latin America in this century (Mexico 1910, Bolivia 1952–53, Cuba 1959–61), and in the case of Brazil this mobilisation broke the bounds of the institutions of labour control, and so posed a direct threat to the social and economic bases of the authoritarian capitalist State; more particularly it threatened the political capability of underpinning the articulation of modes of production, and threatened as well the political base of the landowners, the class which continued in charge of the State, and especially of the key liberal institution of State, the legislature. In other words the peasant mobilisation implied the failure of the institutions of 'diffuse' authoritarianism to impose the political and ideological immobilisation of the peasantry and so reproduce the conditions of accumulation in the countryside. Thus while there was no revolutionary programme, the peasant mobilisation – in conjunction with the struggle in the cities – threatened the social bases of the State and so provoked a general crisis of the State. Whatever the political solution to this crisis it would certainly require far greater and more direct control of the peasantry from the centre, i.e. by central State apparatuses. For Gramsci such a general crisis

> occurs either because the ruling class has failed in some major political undertaking for which it has requested, or forcibly extracted, the consent of the broad masses – or because huge masses (especially of peasantry and petit bourgeois intellectuals) have passed suddenly from a state of political passivity to a certain activity, and put forward demands which, taken together, albeit not organically formulated, add up to a revolution (Gramsci 1973).

The passage from class conciliation to open class struggle logically leads to antagonism between fractions of the dominant classes, which is expressed at the level of State in the conflict between legislature and executive. At the same time the political practices of different classes become detached from the organisations of the traditional political parties, and these parties can no longer channel political demands into the representative institutions of the State. These institutions had first been established within the paradoxical matrix of an authoritarian State which grew through integration into the market and under the aegis of liberal capitalist States of the centre. The export sector earned the resources to build bureaucracy and army, but the influence of liberal ideology originally favoured the flowering of liberal institutions of State. The continuing importance of the export sector to the economy had assured the recurrent survival of these institutions and the resilience of the ideology. But now the contradictions of dependent capitalist growth could no longer be resolved through the liberal institutions of State; the 'civil society', in Gramsci's terms, could no longer maintain the minimum of consensus and consent necessary to an ideological domination of society. More specifically the ideological apparatuses of the State (Poulantzas 1974b) such as *sindicatos*, church, universities; and the political institutions (both 'representative' in the cities and 'controlling' in the countryside) could no longer guarantee the conditions for expanded accumulation. At this moment the different dominant classes and class fractions ('economic-corporate' groups) looked to the one institution capable of guaranteeing their general interests – the army. This movement resulted in 'the fusion of an entire class under a single leadership, which alone is held to be capable of solving an overriding problem of its existence' (Gramsci 1973). Thus 1964 was much more than a mere military intervention. In concrete terms, the fear of an imminent disintegration of the dominant classes in the face of the dissolution of 'civil society' determined a reversion to the moment of pure force – which is the moment of the emergence of the political State.

This crisis of 1964 was a general crisis of the State which, in Gramsci's language, includes on the one hand a 'crisis of representation' (an ideological crisis) and on the other a 'hegemonic crisis'. The ideological crisis refers to the detachment of the 'broad masses' from the traditional parties, and from the corporativist mechanisms of cooptation and integration; the 'hegemonic crisis' to the disintegration of bourgeois hegemony. 'The bourgeois class poses itself as an organism in continual movement, capable of absorbing the entire society, assimilating it to its own cultural and economic level.' At the moment of crisis the bourgeois class is 'saturated'; 'not only does it not expand, it starts to disintegrate; it not only does not assimilate new elements, it loses part of itself' (Gramsci 1973). As in the Brazilian case the bourgeoisie was never itself hegemonic, and the assimilation was achieved through the direct mediation of the State, it was only at the level of the State that a solution could be found to the crisis – assuming that relations of accumulation were to remain the same. Thus the crisis did not represent a mere institutional impasse within the form of the liberal State, but the final crisis of transition from a diffuse system of authoritarian institutions to a highly centralised authoritarian capitalist State, and insofar as it was peasant mobilisation which provoked this crisis, the solution can be seen metaphorically as the imposition of authoritarianism on the city by the countryside.

There had been previous attempts at imposing such a solution, as during the Estado Novo, and many authors now recognise the continuity of the political tradition of authoritarianism between the Estado Novo and the contemporary period (e.g. Skidmore 1967; Stepan 1971, 1973). But by 1964 the formation of the State had been far further determined by the marked dominance of the capitalist mode of production, and an extended articulation of modes of production. The State was literally formed by its functions in the expansion of the sub-capitalist environment; it grew in terms of the dialectic between its own apparatuses and landed interests on the periphery. The traditional landed class was gradually losing power within the overall process, and its decline would be clear in the composition of the power bloc after 1964. But the historical formation of the State absorbed its 'political legacy' at the very moment of that legacy's historical decline: 'when the specific "will" of this stratum [the landowners] coincides with the will and immediate interests of the ruling class . . . then its "military strength" at once reveals itself, so that sometimes, when organised, it lays down the law to the ruling class, at least as far as the "form" of the solution is concerned, if not the content' (Gramsci 1973).

Before examining the form of the solution, two reservations must be made about the broad relationships that are drawn here between the 'ethical–political aspect of politics and the theory of hegemony and consent', on the one hand, and, on the other, 'the aspect of force and economics'. In the first place there is no intention of making the question of articulation of capitalism with subordinate modes of production uniquely determining of events at the political level. A full analysis of the crisis of 1964 must also depart from a series of propositions regarding the penetration of monopoly capital into Brazil's manufacturing sector, the social results this brings about in terms of employment, distribution of income, real wage levels and concentration in industry, and the political response to these social and economic effects. Indeed if it is agreed that the liberal institutions of the State could not contain the contradictions of dependent capitalist growth, then these acute contradictions are precisely those of the 'monopoly capital' period (Ianni 1968). So an analysis departing from the social and economic bases of the authoritarian State cannot be an alternative to explanations deriving from the penetration of monopoly capital, but only a 'historical' complement and corrective to analyses which themselves attempt to explain the crisis uniquely in terms of contemporary development. The impact of monopoly capital and the growth of the urban proletariat, among other things, make further compromises and class conciliation at the level of the State impossible: in Gramsci's language, once again, this State is flawed by an 'insuperable organic deficiency', which is its dependence on the centres of world capitalism or, better, its foundation in internal relations of production which but reflect autonomous capitalist growth in those centres (just as this deficiency created the necessity for the hegemonising alliance which was achieved through the State, so it now causes the collapse of that alliance and the institutions of civil society). In the second place, there is evidently no intention of interpreting the liberal institutions of the civil society as the only means of achieving consensus and consent within the society. On the contrary, while they played an important complementary role in the urban context, the whole analysis has emphasised that bureaucracy and law, and sometimes violence, are the principal mediators of the authoritarian State-in-formation, and these mediators were greatly reinforced and expanded in the post-1964 period.

In this sense it is a mistake to place too much emphasis on the military nature of the State in Brazil after 1964. Evidently there is not only centralisation but militarisation, as more repression and control is exercised from the centre by police and army apparatuses. But not all political power is surrendered to the military, and the military is not, as is often argued, the new 'political party' of the bourgeoisie, or the 'instrument' of monopoly capital or the multi-national corporations. It is the institution which in the first instance directs and oversees the construction of a new form of regime in Brazil, but this is to some degree a historical accident; in other words this role might have been played by some other institution in a different conjuncture, or even by some charismatic leader. What does define the new form of regime politically is the central place and function of bureaucracy, both administrative and legal. But once again this is not because the military cannot dismantle the unwieldy bureaucratic apparatus it inherits, and so must accommodate unwillingly a cluster of more or less autonomous civilian groups which have their economic and corporate interests protected by this bureaucracy. The idea, proposed by Schmitter (1971) that the military is faced with an autonomous political class incorporated into the administrative apparatus of the State, which sabotages the military command structures and absorbs and diverts military objectives, is somewhat absurd. For this bureaucracy is plainly both civil and military and *as a whole* is central to the new form of regime as a *result* of the general crisis of the State, and the political functions, now thrown further into relief, which it fulfils. 'In examining such phenomena [as the crisis of State and its consequences] people usually neglect to give due importance to the bureaucratic element, both civil and military' (Gramsci 1973).

The basic role of bureaucracy after 1964 is to substitute administration for politics or, more specifically, to substitute administrative 'consensus' for political consensus. It will immediately be apparent that in real terms this represents not a new departure but an accentuation of its traditional role. In the absence of the institutions of civil society which contributed to create consensus and consent, and in the absence of more open political mechanisms for interest representation, the bureaucracy now seeks to represent and reconcile economic and corporate interests through its traditional mechanisms of 'bureaucratic entrepreneurship' and the manipulation of the law. Bureaucracy is the new and embracing representative social category, and consequently the key to the expression and exercise of political power in the new form of regime. Regarding its representative functions the earlier discussion on the more direct links being forged between capital and economic interest groups, and Federal State bureaucracy, should now be recalled: these links result not only from the active interest of capital in seeking representation, but from the changing composition of the administration, with the creation and concentration of resources in regional and sectoral agencies which are not only available to act in the interest of capital, but themselves actively encourage private enterprise and seek to advance accumulation in the private sector. The growth of these agencies it was noted in Chapter 7 greatly reduces the areas of autonomy and the degree of executive power of the local states, and in this connection centralisation and bureaucratisation are now consonant and mutually complementary processes.

The legal apparatus of the bureaucracy and the manipulation of the law are also more necessary than ever to the accommodation and reconciliation of conflicting economic and political interests. That manipulation examined as a particular case of bureaucratic intervention on the frontier now pervades every aspect of the political domination of the society, and so demonstrates the paramountcy of the 'political State': 'the law is the repressive and negative aspect of the entire, positive civilising activity undertaken by the State' (Gramsci 1973). The pervasive incidence of the law is so pronounced that convincing analyses have been made of the political development of the new form of regime in terms of the Weberian category of 'institutional charisma' (Klein 1976, with its implications of 'irrational bureaucracy'): there are no constituted or constitutional norms for the drafting, promulgation and regulation of law; on the contrary law is 'invented' on a continuing and immediatist ad hoc basis (and often ex post facto basis) in order to meet the social and political requirements of the moment, which often mean the immediate demands of economic interests and the immediate imperatives of capital accumulation. The law is not predictable but capricious; and its *voltes-face* and irregularities reflect the irrationalities of accumulation within authoritarian capitalism.

Thus the political State is bureaucratic by definition, military by chance. The bureaucracy is the most dynamic element of the centralised State, and, in conditions where most mobilisation is actively discouraged, promotes a surrogate mobilisation around the bureaucratic goals of national economic growth and national security. In this role the bureaucracy is the carrier of the principal elements of the new ideology of the new form of regime which are, precisely, economic development and national security. Ever since the 1920s and the *tenentista* movement, elements of this 'alternative' to liberal ideology had been present in the political arena, manifested in calls for a more centralised and stronger State and in expressions of economic nationalism (Jaguaribe 1969) (this refraction of bourgeois ideology had largely been transmitted historically by the petit bourgeoisie). Now these two principal elements are combined in a highly pragmatic form of nationalism which is in fact as much an 'imperialist' ideology as laissez-faire was 'neo-colonial'. Given the demise of many of the old ideological apparatuses of the State, the new ideology is practised within the bureaucracy: planning, regional and sectoral agencies practise the ideology of development, while the military apparatus and specialist police and investigatory organs practise the ideology of national security; and, at least over a considerable period, the two practices were consciously linked, in the justification of widespread repression as necessary to achieve more rapid rates of growth. In the theory, the distinction between military rule and civilian rule in political science is itself an ideological one; and in practice the civilian and military bureaucracies in Brazil after 1964 are not to be and cannot be distinguished.

The political State and the frontier

It has often been observed that despite the centralisation of the State apparatus and the concentration of power in the executive, the military governments in Brazil after 1964 have conspicuously failed to carry out social and economic reforms in the countryside. The failure to implement such reforms in the past has always been explained by the lack of ideological, instrumental and political conditions – meaning principally that the land-owning class was both an important participant in the power bloc, and over long periods, the class-in-charge of the State. Today it is not one or the other, and the new form of regime certainly commands the political instruments to effect such reforms, but yet does nothing. In this respect the different governments since 1964 did not differ very much in their basic orientation. It should be noted that the 'revolutionary' regime already disposed of a basic law (Estatuto do Trabalhador Rural, Law 4214 of March 1963) sufficiently broad ranging to alter in depth the labour process and land-holding in the countryside (Marcondes 1963); that in three years the Castello Branco government did almost nothing to implement the Estatuto da Terra (Law 4504 of November 1964), the full force of its first article being employed only in a very few isolated cases (though these themselves indicate that it could have been applied more widely) (Ianni 1971); that when Medici returned to the agrarian problem in 1970 he was predominantly preoccupied with credit, prices and technical assistance, intending to raise productivity but with no thought of the 'structural changes' which would mean an agrarian reform. Certainly much of the current political discourse invokes the idea of agrarian reform, and the massive State land agency is still called The National Institute of Colonisation and Agrarian Reform, but even though the reform laws exist they are never implemented. In every region of Brazil the local branches of the land agency continue to calculate and correct the criteria for reform, yet remain captive to local landed interests: 'bureaucracies resemble traditional structures and yet are modern. They can therefore fit themselves to the forms of pseudo-reform' (Feit 1973). This is the classic case of what has been termed in the analysis 'bureaucratic inertia'.

Instead of carrying out a reform in the countryside, the new form of regime in Brazil continues to pursue a pattern of horizontal growth, which takes the form of continuing expansion of the frontier, extended reproduction of the sub-capitalist environment, and an acceleration of the process through incentives to capital and modifications in the form of articulation between the dominant capitalism and the rest of the economic environment. J. F. Normano wrote in 1935 that Brazilian development was 'a process of putting an economic substance in the political arena, of bringing the economic nearer the political frontier', and in his preface left it quite clear that this process was deliberate: 'the aim was and is the same – the extension of the economic territory'. In the light of his words it is evident that the new form of regime in Brazil is pursuing a highly traditional pattern of growth in its commitment to the extension of the frontier, rather than to vertical reforms within already occupied and developed regions.

This conclusion has become inescapable since the massive State promoted expansion of the frontier in the Amazon:

The government's position reveals the official determination to

> develop by means of the geographical expansion of the economic frontier, instead of by structural and technological reforms in the regions which have already been integrated into the economic system. This implies, obviously, that the development of Amazônia takes priority over that of the North-East, given that the development of the two regions involves different premises and divergent economic strategic perspectives (Sá 1970).

The results of this Amazon priority have already been investigated: a costly road-building programme centred initially on the Transamazônica; new State agencies equipped with sophisticated information services, and commanding large proportions of the national budget, specifically designed to promote the development of the region; a range of fiscal and credit incentives to attract private enterprise and occupy the region as fast as possible; the invitation to all capital, including especially foreign capital, to enter the region and explore its natural resource base and mineral deposits of every kind. And even within a few short years, the impact on the region seems to fulfil Normano's 'prophecy' of forty years ago:

> Brazil is supplied with an immense reserve of land for expansion, with an enormous potential market for an industrialised country. Not a territorial political expansion but a populationist one is actual here. No search for new markets, for new territories, for raw materials is necessary. Everything exists and waits for new *bandeirantes* equipped with initiative, capital and modern methods. It is a quest for a new move of the frontier (Normano 1935).

The frontier continues to expand. New methods are certainly applied, but the pattern of growth is traditional, precisely through the expansion of the frontier. The land tenure structure of the occupied regions is not reformed, but on the other hand the pattern of horizontal growth continues. To some degree such continuity can be explained in purely political terms as conducive to a certain social and political stability. The horizontal growth can be seen as 'absorbing' structural contradictions in the countryside, and therefore as permanently postponing the necessity for structural or vertical reforms: as Ianni says of the new form of regime, 'the "pragmatism" of the Government revealed very clear objectives. It wanted economic development in conditions of social and political stability. It was a question of consolidating and perfecting the status quo' (Ianni 1971). But as far as development in the countryside goes, it should already be apparent that vertical reform and lateral growth are not *economic* alternatives. In economic terms the 'bureaucratic inefficacy' and 'political inertia', i.e. the absence of structural reforms, is necessary to the continuing presence of horizontal growth and lateral accumulation. It is precisely the 'unreformed' countryside, latifundio and minifundio, which provides the general conditions for this lateral accumulation – which means economically a continuing high rate of exploitation of peasant labour and a continuing rapid rate of appropriation and transfer of value from countryside to city. In simple language, the motor of economic growth in the countryside continues to be the expansion of the frontier, and accumulation within non-capitalist relations of appropriation (or 'mixed' relations of appropriation).

Since its beginnings the authoritarian State in Brazil has practised permanent political intervention in order to promote the conditions of accumulation in the countryside, and this continues to be the case today. Given the massive availability of land and labour on the frontier the rapid combination of the two continues to provide the best possibility for rapid accumulation, and therefore it should be no surprise that the new form of regime (of the authoritarian form of the capitalist type of State) in Brazil continues to promote this traditional pattern of growth. Accumulation is the key to understanding both the expansion of the frontier and the intervention of the State in this process. This is proven more completely in the contemporary context than in the past. The State now intervenes ahead of the arrival of the frontier – opening access through the construction of roads, transporting labour through colonisation projects, providing capital through incentive schemes – precisely in order to accelerate the accumulation which takes place on the frontier. Since the 1930s the bureaucratic apparatuses and agencies have followed the pioneers – peasants and entrepreneurs – to the frontiers, and by intervening in the occupation of the land, the definition of private property and the marketing of produce have contributed to the appropriation of the peasant surplus: in the contemporary context the range of 'economic activity' of the bureaucracy is extended to include the first stages of the process, which are the movement to the frontier and the 'spontaneous' occupation of the land.

But the primary political functions of the State in the process of frontier expansion and accumulation have not changed, but simply become more complete, in the same way that the form of State in Brazil has not changed since the 1930s, but has simply become more politically consistent. During this period this State has continued to be capitalist and authoritarian, as defined by its social and economic bases in the countryside and by its primary political functions as expressed so clearly in its intervention on the frontiers. Over certain periods an evident discrepancy or disjunction has existed between the economic bases and the political functions of the State (its 'mediations'), and the political institutions of the State, which, due partly to the resilience of liberal ideology and partly to the necessity of maintaining its political functions in the face of a mass political population in the cities, remained 'liberal' in form. However the construction of new State institutions in the years after 1964 means that today the form of regime of the authoritarian capitalist State *corresponds* more closely than ever before to its social and economic bases, and to its primary political tasks. In this sense, it is politically more consistent. In the same way, the content of State intervention on the frontier has not changed but continues to be determined by, and defined in relation to, the process of accumulation in the countryside. Yet today the greater political consistency of the State, the centralisation of its apparatuses and concentration of power in the executive, and the greater ideological coherence of its economic policy initiatives, make this intervention more complete than at any other time during the past fifty years.

Bibliography

Official Publications

ACARPA, 1966, 1967, 1968 (Associação de Crédito e Assistência Rural do Paraná) – *Realidade Rural* (statistical reports on individual municipalities, especially Francisco Beltrão, Pato Branco and Dois Vizinhos).

Amaral Batista, E. & Goes Correa, C. S., 1975, *Aspectos territoriais e demográficos da Amazônia Legal.* SUDAM – Divisão de Estatística.

BASA (Banco de Amazônia) & Universidade do Pará, 1967, Desenvolvimento econômico da Amazônia.

Cano de Arruda, R., 1967, *Reforma Agrária* – (Instituto Interamericano de Ciencias Agrícolas da OEA-IICA-Projeto 206 do Programa de Cooperação Técnica da OEA) Porto Alegre.

CIBRAZEM (Companhia Brasileira de Armazenagem), 1967, Pesquisa básica para um programa global de armazenagem intermediária. SPL (Serviços de Planejamento).

CIDA (Comitê Interamericano de desenvolvimento agrário), 1966, *A posse e uso da terra e o desenvolvimento socio-econômico do setor agrícola no Brasil.* Washington D.C.

CNBB, 1976, Comunicação Pastoral ao Povo de Deus, separata do *Comunicado Mensal da CNBB*, CNBB (Conferência Nacional dos Bispos do Brasil). Rio de Janeiro.

CODEPAR (Companhia de Desenvolvimento do Paraná), 1964, *O Paraná e a Economia Madeireira.*

CODEPAR & Sociedade SERETE de estudos e projetos Ltda, São Paulo, 1965, *Desenvolvimento econômico das regiões oeste e sudoeste do Paraná.* December.

Comissão Interestadual da Bacia Paraná-Uruguai, 1964, *Alguns aspectos de desenvolvimento regional – Plano de industrialização regional* (Alguns aspectos da economia de Mato Grosso).

Comissão Pastoral da Terra, 1976, *Boletim*, Nov./Dec.

Conselho Nacional de Geografia, and IBGE (Instituto Brasileiro de Geografia e Estatística), 1963, *Geografia do Brasil* Vol. IV. *Grande Região Sul.* Rio de Janeiro.

Esteves, J., 1976, *Atuação do Ministério do Interior através da SUDAM e SUFRAMA* (Discurso no Senado Federal, 19 Abril). Centro Gráfico Senado Federal. Brasília.

IBGE (Instituto Brasileiro de Geografia e Estatística), 1959. *Enciclopédia dos municípios brasileiros* XXXI. Rio de Janeiro.

1960. *Anuário Estatístico do Brasil.*

1970. *Sinopse preliminar do censo demográfico.* Pará.

1970. *Sinopse preliminar do censo demográfico.* Mato Grosso.

1970. *Sinopse preliminar do censo demográfico.* Paraná.

1970. *Dados preliminares gerais do censo agropecuário.* Região Sul.

IBRA (Instituto Brasileiro de Reforma Agrária), 1967a. *Cadastro.* Paraná.

1967b. *A estrutura agrária brasileira.*

1971. *Cadastro.* Paraná.

IGRA (Instituto Gaúcho de Reforma Agrária), 1966. *Bases de diretrizes para um programa estadual de reforma agrária.* Porto Alegre.

INCRA (Instituto Nacional de Colonização e Reforma Agrária), 1970, Departamento de operações e projetos (Distrito de Terras do Sul de Mato Grosso). *Projeto Iguatemí.*

1971, Coordenadoria Regional do Rio Grande do Sul. *Levantamento de recursos naturais, socio-econômicos e institucionais do Estado.* INCRA-IICA-Governo do Estado.

1974, *Estatísticas cadastrais.* Sistema Nacional de Cadastro Rural. Departamento de Cadastro e Tributação – Programa Geral de Estatísticas. Recadastramento 1972. Brasília D.F.

1975, *Realizações e metas.* Coordenadoria Geral. Brasília, D.F.

INDA (Instituto Nacional de Desenvolvimento Agrário) – AMSOP-GETSOP, 1969, *Pre-diagnóstico socio-econômico do sud-oeste do Paraná.*

ITERPA Lawyers & Advisers, 1976, Articles and notes on legal history of land in Pará, and the present disputes. *Revista do ITERPA* No. 1 Jan; No. 2 April; No. 3 July.

Lourdello de Mello, D., 1971, *O município na organização nacional.* IBAM (Instituto Brasileiro de Administração dos Municípios) Rio de Janeiro.

Magno de Carvalho, J. A. & Mello Moreira, M., 1976, *Migrações internas na região norte* Vols. I & II. SUDAM – Departamento de Recursos Humanos convênio SUDAM-CEDEPLAR (Centro de desenvolvimento e planejamento regional). Belém.

Ministério de Planejamento e Coordenação Econômica, 1966, *Demografia, diagnóstico preliminar.* Rio de Janeiro.

Mozart, C.; Cramer, E. R.; Dorival, V. S., 1956, *Estudos sobre trigo* (SAPS – Serviço de Alimentação da Previdência Social).

Relatório Geral do Governo Moisés Lupion 1956–1960 (Curitiba 1961).

Revista Bancária Brasileira, 1956, Núcleos coloniais. No. 281. May.

Silva, M. J. O., 1974, *A sociedade amazônica e o problema social de ocupação e sub-ocupação.* SUDAM – Departamento de Recursos Humanos.

Souza Melo, A. & Souza Campos, J., 1976, *Algumas considerações sobre os problemas socio-econômicos enfrentados pelos migrantes na area do polo Carajás.* SUDAM DOCUMENTA (Documentos Amazônicos 6.1/4).

SUDAM, 1973, *Elementos para o aperfeiçoamento da ação planejada do poder público na Amazônia.* Belém.

1976a, *Incentivos financieros e fiscais: regulamentação.* SUDAM Divisão de Documentação. Belém.

1976b, *POLAMAZÔNIA (Programa de Polos Agropecuários e Agrominerais da Amazônia) Sintese.* SUDAM – Coordenação e Informação – Divisão de Documentação.

Other primary sources

Banaji, J., n.d., Harris on the peasantry. Oxford International Socialists.

Brandão Lopez, J. & Patarra, N. L., 1974, Redistribução regional e rural–urbana da população brasileira. CEBRAP. São Paulo. mimeo.

Bunker, S. G., 1978, Class, status, and the small farmer: rural development programs and the advance of capitalism in Uganda and Brazil. NAEA (Núcleo de Altos Estudos Amazônicos) Universidade Federal do Pará.

Chalmers, D. A., 1970, Political groups and authority in Brazil. Columbia University, New York.

CODEPAR (Companhia de Desenvolvimento do Paraná), 1963, Agricultura de subsistência no Paraná. Curitiba.

1965, Estudo sobre o estatuto da terra e seus possíveis reflexos no Paraná. Curitiba.

Comninos, C., 1971. Alguns aspectos populacionais do Paraná. Curitiba.

Correio da Manhã (Rio de Janeiro), editions of Sept.–Oct. 1957.

Costa de Albuquerque, J., 1970, *Relatório*: Enfoque da problemática fundiária existente na faixa de fronteira dos Estados do Paraná e Santa Catariná, cuja região encontra-se jurisdicionada ao DFZ-01. Cascavél. Oct.

Davis, S. H. & Mathews, R. O., 1977, *The Geological Imperative: anthropology and development in the Amazon basin of South America*. Occasional papers No. 5. Program in Comparative Culture. University of California at Irvine.

DGTC (Departamento de Geografia, Terras, e Colonização – Paraná), 1966, *Relatório* – General Gaspar Peixoto Costa. Curitiba.

Diário da Tarde (Curitiba), editions April–Oct. 1957.

DTC (Divisão de Terras e Colonização, Paraná), 1933, Relatório – Othon Mader. 1948, Relatório.

DTCC (Departamento de Terras, Colonização e Cooperativismo do Governo de Estado do Pará), 1974, Levantamento preliminar das glebas que compôem o bloco FLORESTA

Elfes, A., 1970, Estudos agro-econômico e social. INDA – Instituto Nacional de Desenvolvimento Agrário – Delegacia Regional do Paraná.

Erven, B. L., 1969, Uma análise econômica de uso de crédito rural e de problemas de política creditícia no estado do Rio Grande do Sul. IEPE (Instituto de Estudos e Pesquisas Econômicos) – Universidade Federal do Rio Grande do Sul.

Estado de São Paulo, especially editions of May 1974 to December 1975, and, in particular, the series 'Amazônia: dez anos de colonizacão' by Lucio Flavio Pinto *et al.* 3–11 Nov. 1975.

Estado do Paraná (Curitiba), editions April–Oct. 1957.

Esterci, N., 1972, O mito da democracia no país das Bandeiras – Análise dos discursos sobre colonizacão e migração no Estado Novo. Pesquisa Antropológica. No. 18 Divisão de Antropologia, Museu Nacional. Rio de Janeiro.

Feder, E., 1973, Dependency and the agricultural problem in L.A. Workshop on Dependency in Latin America. Amsterdam. CEDLA: 19–21 Nov.

Fôlha de São Paulo, especially editions of May 1974 to December 1975.

Foweraker, J. W., 1971, The frontier in the south-west of Paraná. B.Phil. Oxford.

1974, Political conflict on the frontier. D.Phil. Oxford.

1975, The social bases of conservatism in Brazil.
1975, The politics of colonization in the south of Mato Grosso.
1975, The State and economic development: the impact on the periphery.
Genshow, F. & Alarico da Cunha, Jr, 1957, *Conceito da colonização*. Tese aprovada no IV Congresso Nacional de Municípios. Rio de Janeiro.
GETSOP (Grupo Executivo das Terras Sud-oeste do Paraná), 1966, *Relatório* – Ildephonso G. de Oliveira, 27 January.
Glock, O., 1976, Capitalist development, rent and agrarian reform: the Chilean Case. M.A. thesis, University of Essex, Jan.
Gomes, B. B. (IBRA attorney), 1969, Relatório re. Processos do SPU ao chefe DFZ-01. Cascavél, 8 Nov.
Hirst, P. Q., 1976, Can there be a peasant mode of production?
Hoffman, R., 1967, Contribuição a análise da distribuição da posse de terra no Brasil. Escola Superior Agrícola Luis de Queiroz. Piricicaba.
1972, Tendências da distribuição da renda no Brasil e sua relações com o desenvolvimento econômico. Luis de Queiroz, Serie Estudos No. 13. Piricicaba.
Ianni, O., 1977, *A luta pela terra* CEBRAP – Programa de População – Pesquisa Nacional sobre Reprodução Humana, São Paulo. March.
IEPE (Instituto de Estudos e Pesquisas Econômicos), 1971, Formação de capital e mudanças tecnológicas ao nivel de empresas rurais – Lajeado, Carazinho, e Não-me-Toque (RGS). Eli de Moraes Souza, *et al.* Universidade Federal do Rio Grande do Sul.
Isto é: No. 9 of January 1977.
Jordão Neto, A., 1967, Migrações. São Paulo.
Katzman, M. T., 1977a, The geopolitical significance of Brazilian frontier expansion. Presented to Latin American Studies Association annual conference. Houston, November.
Keller, F. I. V., n.d., Relatório do 'Survey' realizado no Brazil Centro-Oeste. Projeto Estudo Comparativo do Desenvolvimento Regional. Programa de Pos-Graduação em Antropologia Social, Divisão de Antropologia. Museu Nacional. Rio de Janeiro.
1973, O homem da frente de expansão: permanência, mudança e conflito. Divisão de Antropologia, Museu Nacional.
Lacerda, C., 1957, Speech in the *Camara Federal*. October.
Lamounier, B., 1974, Ideology and authoritarian regimes: theoretical perspectives and a study of the Brazilian case. Ph.D. thesis, University of California, Los Angeles.
O Liberal, especially editions of November 1976 to July 1977. Belém.
Machado Neto, C., 1971, Conceitos e dados sobre colonização e reforma agrária. Coordinador do INCRA-MG (CR-06) Universidade Federal de Viçosa, July.
Mader, O., 1957, *Dois discursos pronunciados no Senado Federal em 6 e 9 de dezembre pelo Senador Othon Mader*.
Martins, J. de Souza, 1969a, Modernização e problema agrário no estado de São Paulo. Instituto de Estudos Brasileiros. Universidade de São Paulo.
1971, Frente pioneira: contribuição para uma caracterização sociológica. Comunicação apresentada sob o patrocínio da Fundação de Amparo a Pesquisa de São

Paulo, a XXIII Reunião Anual da Sociedade Brasileira para o Progresso da Ciência. Curitiba, 4–10 July.

1972, Adoção de prácticas agrícolas e tensões sociais. Faculdade de Ciencias Sociais, Universidade de São Paulo.

Mendes, A. *et al.*, 1974, A invenção da Amazônia. NAEA (Núcleo de Altos Estudos Amazônicos). Belém.

Messias Junquiera, 1966, As terras públicas no estatuto da terra. July.

Ministerio da Agricultura, Ministerio do Planejamento e Coordenação Geral, Ministerio da Fazenda, Ministerio do Interior, IBRA, Confederação Nacional da Agricultura, Confederação Nacional dos Trabalhadores da Agricultura, 1968, *Reforma agrária* relatório do grupo de trabalho inter-ministerial criado pelo Decreto 63,250 de 18 Set. 1968.

Molina, M. I. G., 1970, Análise sociológica da migração dos parcelarios do projeto Iguatemí. Tese de doutoramento apresentada a Escola Superior da Agricultura Luis de Queiroz. Piricicaba.

Monteiro, B., 1962, *Projeto de reforma agrária no Pará.*

1963, *Reforma agrária na lei o na marra.*

Moran, E. F., 1975, Pioneer farmers of the Transamazonian highway: adaptation and agricultural production in the lowland tropics. Ph.D. dissertation, University of Florida, Gainesville.

Moura Castro, C., 1975, Ecologia – a redescoberta da pólvora. FIPAM doc. no. 666 NAEA (Núcleo de Altos Estudos Amazônicos) Universidade Federal do Pará.

Muller, G. & Brandão Lopez, J., 1975, *Amazônia: desenvolvimento socio-econômico.* Vol. I. CEBRAP.

Muller, G., Vasconcellos, T. M. A., Cardoso, F. H., & Brandão Lopez, J., 1975, *Amazônia: políticas de população.* Vol. II. CEBRAP. Programa PISPAL (Programa de investigaciones sociales sobre problemas de población relevantes para políticas de población en América Latina).

Mundo: Ano. I no. I July–September 1975.

NAEA (Núcleo de Altos Estudos Amazônicos), 1974, A Amazônia no processo de integração nacional. Equipe FIPAM (Programa International de Formação de Especialistas em Desenvolvimento de Areas Amazônicas). Universidade Federal do Pará.

1975, Colonização dirigida na Amazônia. FIPAM-NAEA. Universidade Federal do Pará. Belém.

Nicholls, W., n.d., The agricultural frontier in modern Brazilian history: the state of Paraná, 1920–65.

Oberg, K. & Jabine, T., 1957, *Toledo: a municipio on the western frontier in the state of Paraná.* USOM Brazil. Rio de Janeiro.

Oliveira, F., 1972a, A economia brasileira: notas para uma revisão teórica. CEBRAP (Centro Brasileiro de Análise e Planejamento). São Paulo.

Oliveira, F. & Reichstul, H.-P., 1972, Mudanças na divisão interregional do trabalho no Brasil. CEBRAP. São Paulo.

Palma Arruda, H., n.d., *Colonização na Amazônia brasileira.* INCRA – Departamento de Projetos e Operações.

Palmeira, M. G. S., 1970, Latifundium et capitalisme au Brésil: lecture critique d'un débat. Thèse de 3e cycle presentée à la Faculté des Lettres et Sciences Humaines de l'Université de Paris.

Pastore, C., 1968, A resposta da produção agrícola aos preços no Brasil. Universidade de São Paulo.

Pereira de Queiroz, M. I., 1957, *La 'Guerre Sainte' au Brésil: le mouvement messianique du 'Contestado'*. Boletim no. 187, Faculdade de Filosofia, Ciências e Letras, Universidade de São Paulo.

Pompermaier, M., 1979, The State and the frontier in Brazil. Ph.D. thesis. Stanford.

Praefeitura Municipal de Palotina, 1971, Relatório – ao Coronel Floriano Aguilar Chagas DD Secretário Executivo da Comissão Especial da Faixa da Fronteira. Brasília D.F. 23 August.

Procuradoria Geral da República, 1968, Pedido (to President of STF) August.

Realidade: supplement to July 1970 edition (containing studies from Fundação Getúlio Vargas).

Relatórios, n.d., by the directors of the DGTC (Departamento de Geografia, Terras e Colonização) of Paraná – (Collection of the DGTC in Curitiba).

Relatório, 1968, on the land tenure situation in the south-west of Paraná, by Oswaldo Aranha, and fellow agronomists within the area.

1970a, Addressed to General Mario Carneiro Portes, Secretário de Estado, Segurança Pública do Estado do Paraná – from Oscar Pacheco dos Santos, Delegado Chefe da 5 Subdivisão Policial de Pato Branco. 23 January.

1970b, Addressed to Dr Cicero do Amaral Viana, Superintendente dos Serviços Policiais, Porto Alegre – from Antonio Dufech, Delegado da Policia de Porto Alegre, Rio Grande do Sul. 3 March.

Sá, J. I., 1968, Utilização de mão-de-obra e níveis de renda em pequenas propriedades rurais – Santa Rosa – RGS. IEPE – Universidade Federal do Rio Grande do Sul.

Schmink, M., 1977, Frontier expansion and land conflicts in the Brazilian Amazon: contradictions in policy and process. Presented at meeting of American Anthropological Association. Houston, Dec.

Schmink, M. & Wood, C., 1978, Blaming the victim: small farmer production in an Amazon colonization project. Interciencia Association Symposia. Washington.

Seffer, E., 1976, Address to the Assembléia Legislativa do Pará. INCRA – Coordenadoria Regional. Dec.

Shigueru, H., 1972, Relatório sobre a fazenda Santa Rita de Cássia. DFZ-01. Internal. INCRA. Cascavél.

Singer, P., 1972, O 'Milagre Brasileiro': causas e consequencias. CEBRAP.

Smith, G. W., 1965, Agricultural marketing and economic development: a Brazilian case study. Ph.D. thesis, University of Harvard. October.

Tourinho, Gen. L. C., 1968, *Relatório das atividades da interventoria no período de 23 julho a 31 de dezembro 1968*. Boletim do IBRA No. 206, Suplemento. 31 December. Rio de Janeiro.

Tribuna da Imprensa (Rio de Janeiro) editions of 1957.

Tribuna do Paraná (Curitiba), editions April–Oct. 1957.

Ultima Hora (Rio de Janeiro), editions of 1957.

Velho, O. G., 1973, Modes of capitalist development, peasantry and the moving frontier. Ph.D. thesis, University of Manchester.

Westphalen, C. M.; Pinheiro Machado, B.; Balhana, A. P., 1968, *Nota prévia ao estudo da ocupação da terra no Paraná*. Boletim da Universidade Federal do Paraná, Conselho de Pesquisas – Departamento da História No. 7.

Whitehead, L., 1975, An attempt to re-habilitate the 'State'. Oxford.

Wolf, E., 1967, Reflections on peasant revolutions. Carnegie Seminar on Political and Administrative Development. Indiana University, Bloomington, 3 April.

Secondary sources

Alavi, H., 1965, Peasants and revolution. *Socialist Register*, eds. Miliband & Saville.

1973, Peasant classes and primordial loyalties. *Journal of Peasant Studies* Vol. 1 No. 1. October.

1975, India and the colonial mode of production. *Socialist Register*, eds. Miliband & Saville.

Alves, M. M., 1973, *A grain of mustard seed*. Doubleday Anchor, New York.

Alvez de Souza, S., 1965, Distribuição da população da região centro-oeste – 1960. *Revista Brasileira de Geografia* Ano 27(3).

Amin, S., 1970, La transition au capitalisme péripherique. In *L'Accumulation a l'échelle mondiale*. Anthropos, Paris.

Anderson, C., 1967, *Politics and economic change in Latin America*. Van Nostrand Reinhold, New York.

Anderson, P., 1974, *Lineages of the absolutist state*. New Left Books, London.

Andrade, M. C., 1963, *A terra e o homem no nordeste*. Brasiliense, Rio de Janeiro.

Averburg, M., 1969, Reflexos da estrutura agrária sobre o processo da comercialização dos produtos agrícolas. *Agricultura subdesenvolvida* – Caminhos brasileiros 2. Vozes, Petrópolis.

Baer, W., 1964, Regional inequality and economic growth in Brazil. *Economic Development and Cultural Change* XII No. 3. Chicago.

1965, *Industrialization and economic development in Brazil*. Richard D. Irwin. Homewood, Illinois.

Baer, W. & Maneschi, A., 1969, Import substitution, stagnation and structural change – an interpretation of the Brazilian case. *Revista Brasileira de Economia*, No. 1. Jan.–Mar.

Balan, J., 1973, Migrações e desenvolvimento capitalista no Brasil: ensaio de interpretação histórico-comparativa. *Estudos* CEBRAP 5.

1974, ed., *Centro e periferia no desenvolvimento brasileiro*. DIFEL São Paulo.

Baldwin, R., 1964. Problems of development in newly settled regions. *Agriculture in Economic Development*, eds. Eicher & Witt, McGraw Hill, New York.

Barbosa, I. C., 1967, Esboço de uma nova divisão regional do Paraná. *Revista Brasileira de Geografia* No. 3.

Bartra, R., 1974, *Estructura agrária y clases sociales en Mexico*. Ediciones ERA Serie Popular. Mexico, D.F.

Becker, B. K., 1968, As migrações internas no Brasil, reflexos de uma organização do espaço desequilibrada. *Revista Brasileira de Geografia* No. 2.

Bergsman, J., 1970, *Brazil: industrialization and trade policies*. Oxford University Press. London.

Bernardes, L. M. C., 1953, O problema das 'frentes pioneiras' no Estado do Paraná. *Revista Brasileira de Geografia* No. 3. July–Sept.

Bernardes, N., 1950, Expansão do povoamento no estado do Paraná. separata *Revista Brasileira de Geografia* No. 4. October–December.

1952, A colonização européia no sul do Brasil. *Boletim Geográfico*. IBGE. No. 109. July.

1961, A velha imigração italiana e sua influência na agricultura e na economia do Brasil. *Boletim Geográfico* No. 19, March–April.

Betancourt, J. S., 1960, *Aspecto demográfico-social da Amazônia brasileira*. Coleção Araújo Lima – Representação da SPVEA. Rio de Janeiro.

Bettelheim, 1972, Theoretical comments. In *Unequal exchange*, ed., Emmanuel A. New Left Books. London.

Bienen, H., 1970, *Tanzania: party transformation and economic development*. Princeton University Press. Princeton.

Blok, A., 1974, *The mafia of a Sicilian village 1860–1960*. Oxford University Press. London.

Blondel, J., 1957, *As condições da vida política no estado do Paraíba*. Fundação Getúlio Vargas. Rio de Janeiro.

Borges, M., 1965. *O golpe em Goiás: História de uma grande traição*. Civilização Brasileira, Rio de Janeiro.

Botelho, H. S., 1976, A política brasileira de integração nacional: efeitos e perspectivas. IPEA. *Boletim Econômico* 3.

Bourne, R., 1978. *Assault on the Amazon*. Gollancz. London.

Bradby, B., 1975, The destruction of natural economy in Peru. *Economy and Society* Vol. 4. May.

Brandão Lopez, J., 1964. *Sociedade industrial no Brazil*. DIFEL. São Paulo.

1966, Transformations in Brazilian political structure. In *New Perspectives of Brazil*, ed. E. N. Baklanoff. Vanderbilt University Press. Nashville.

Candido. A., 1964, *Os parceiros do Rio Bonito*. José Olympio. Rio de Janeiro.

Capstick, M., 1970, *Economics of agriculture*. London.

Carneiro, D., n.d., *Fasmas estruturais da economia do Paraná*. Universidade Federal do Paraná.

Cardoso, F. H., 1961, Tensões sociais no campo e reforma agrária. *Revista Brasileira de Estudos Politicos* No. 12.

1972a, *O modelo político brasileiro*. DIFEL. São Paulo.

1972b, Dependency and development in Latin America. *New Left Review* No. 74. July/Aug.

1975, *Autoritarismo e Democratização*. Paz e Terra. Rio de Janeiro.

Cardoso, F. H. & Faletto, E., 1969, *Desarrollo y dependencia en América Latina*. Siglo XXI. Mexico.

Cardoso, F. H. & Muller, G., 1977, *Amazônia: expansão do capitalismo*. Brasiliense. São Paulo.

Carneiro, F., 1950, *Imigração e colonização*. Faculdade Nacional de Filosophia.

Casanova, P. G., 1970, *Democracy in Mexico*. Oxford University Press. London.

Castello Branco, J., 1951. Planejamento de colonização em Mato Grosso. *Revista do Serviço Público* Ano. 14(4). October.

Castro, Antonio B., 1969, *Sete ensaios sobre a economia brasileira.* Forense. Rio de Janeiro.

Castro, Armando B., 1970, Aspectos da evolução econômica e social no l'este do Mato Grosso. *Caldernos No.* 3. Centro de Estudos Rurais e Urbanos, Universidade de São Paulo.

Chacel, J., 1969. The principal characteristics of the agrarian structure and agrarian production in Brazil. *The Economy of Brazil.* Ed. H. Ellis. University of California. Berkeley.

Chayanov, A. V., 1966, *The theory of peasant economy.* Homewood, Illinois.

Chisholm, M., 1966, *Rural settlement and land use: an essay in location.* Hutchinson. London.

Cline, W. R., 1970, *Economic consequences of a land reform in Brazil.* North-Holland. Amsterdam.

CNRS, 1967, *Les problèmes agraires des Amériques Latines.* Centre National de la Recherche Scientifique. Paris.

Coelho, S., 1971, Mato Grosso. *Estado de São Paulo.*

Coletti, L., 1972, *From Rousseau to Lenin.* New Left Books, London.

Correa, R. L., 1970, O sudoeste paranaense antes da colonização. *Revista Brasileira de Geografia* No. 1. Jan.–March.

Correa Filho, V., 1957, *Ervais do Brasil e ervateiros.* Serviço de Informação Agrícola.

Davis, S., 1977, *Victims of the miracle: development and the Indians of Brazil.* Cambridge University Press. New York.

Dean, W., 1969, *The industrialization of São Paulo 1880–1945.* University of Texas, Austin.

Diaz del Moral, J., 1967, *História de las agitaciones campesinas andaluzas.* Alianza. Madrid.

Diegues Jr, M., 1959, *Land tenure and use in the Brazilian plantation system.* Pan-American Union – Social Science Monograph VII. Washington, D.C.

Dos Santos, T., 1968, Foreign investment and the large enterprise in Latin America: the Brazilian case. In eds. Petras & Zeitlin *Latin America: reform or revolution?* Fawcett World Library. New York.

1970, The structure of dependence. *American Economic Review* Vol. 60 No. 2. May.

Dumont, R., 1957, *Types of rural economy: studies in world agriculture.* Praeger. New York.

Eicher, C. & Witt, L., eds., 1964, *Agriculture in economic development.* McGraw Hill. New York.

Eidt, R., 1971, *Pioneer settlements in North-East Argentina.* University of Wisconsin. Madison.

Ellis, H., ed., 1969, *The economy of Brazil.* University of California. Berkeley.

Engels, E., 1928, *The mark* (An Appendix to *Socialism, utopian and scientific*). New York.

1934a, *Anti-Dühring* (Landmarks of Scientific Socialism). London.

1934b, *Correspondence Marx–Engels 1846–95.* Martin Lawrence. London.

Enjalbert, H., 1967, Réforme agraire et production agricole au Mexique 1910–1965. *Problèmes agraires des Amériques Latines.* CNRS. Paris.

Faco, R., 1965, *Cangaçeiros e fanáticos: gênese e lutas*. Civilização Brasileira. Rio de Janeiro.

Faoro, R., 1975. *Os donos do poder: formação do patronato político brasileiro*. Globo. Porto Alegre.

Fausto, B., 1975. *A revolução de 30*. Brasiliense. São Paulo.

Feder, E., 1971. *The rape of the peasantry*. Doubleday Anchor. New York.

Feit, E., 1973. *The armed bureaucrats*. Houghton Miflin. Boston.

Fernandes, F., 1968. Capitalismo agrário e mudança social. In *Sociedade de clases e desenvolvimento*. Zahar. Rio de Janeiro.

Ferreira Reis, A. C., 1972. *O impacto amazônico na civilização brasileira*. Paralelo. Rio de Janeiro.

Fontana, A., 1960. A colonização do oeste catarinense. *Anuário Brasileiro de Imigração e Colonização* p. 37.

Forman, S., 1971. Disunity and discontent: a study of peasant political movements in Brazil. *Journal of Latin American Studies* Vol. 3 Part 1.

1975. *The Brazilian peasantry*. Columbia University Press. New York.

Forman, S. & Riegelhaupt, J., 1970a. Market place and marketing system – toward a theory of peasant economic integration. *Comparative Studies in Society and History* No. 12.

1970b. Bodo was never Brazilian: economic integration and agricultural development. *Journal of Economic History*. Spring.

Foweraker, J. W., 1978. The contemporary peasantry: class and class practice. In ed. H. Newby, *International Perspectives in Rural Sociology*. John Wiley. Chichester.

Freire, P., 1972. *Pedagogy of the oppressed*. Penguin. Harmondsworth.

Furtado, C., 1963. *The economic growth of Brazil: a survey from colonial to modern times*. University of California. Berkeley.

1964. *Dialéctico de desenvolvimento*. Fundo da Cultura. Rio de Janeiro.

1965. Political obstacles to the economic development of Brazil. In *Obstacles to change in Latin America*, ed. C. Veliz. Oxford University Press. London.

1968. *Um projeto para o Brasil*. Saga. Rio de Janeiro.

1976. *Prefácio a nova economia política*. Paz e Terra. Rio de Janeiro.

Galeski, B., 1972. *Basic concepts of rural sociology*. Manchester University Press.

Galjart, B., 1968. *Itaguai*. Wageningen Centre for Agricultural Publishing and Documentation.

Garavaglia, J. C., ed., 1973. *Modos de producción en América Latina*. Cuadernos de Pasado y Presente 40. Córdoba B.A.

Garcia, F., 1967. O sertão de Itapecería. *Problèmes agraires des Amériques Latines* CNRS. Paris.

Gastão de Alencar, L., 1971. Faixa de fronteira no Paraná é zona prioritária. *Estado do Paraná*. Curitiba 31 October.

Georgescu-Roegen, N., 1960. Economic theory and agrarian economics. *Oxford Economic Papers* No. 12, February.

Gerchunoff, A., 1957. *Los gauchos judíos*. Sudamericana. Buenos Aires.

Germani, G., 1969. *Sociologia de la modernización*. Paidós. Buenos Aires.

Gerschenkron, A., 1962. *Economic backwardness in historical perspective*, pp. 5–30. Harvard University Press. Cambridge, Mass.

Goodman, D., 1978. Expansão de fronteira e colonizacão rural: recente política de desenvolvimento no centro-oeste do Brasil. In eds. Baer, Geiger & Haddad, *Dimensões do desenvolvimento brasileiro*. Campus. Rio de Janeiro.

Graham, D., 1969. Padrões de convergência e divergência do crescimento econômico e das migrações no Brazil 1940–1960. *Revista Brasileira de Economia* No. 3.

Graham, R., 1966. Causes for the abolition of negro slavery in Brazil. Hispanic American Historical Review. May.

Gramsci, A., 1968. The southern question. In *The modern prince and other writings*, ed., L. Marks. International. New York.

1973. Notes on Italian history & State and civil society. In *Selections from the prison notebooks*, eds. Hoare & Smith. Lawrence and Wishart. London.

Griffin, K., 1968. Coffee and the economic development of Colombia. *Bulletin of the Oxford University Institute of Economics and Statistics* Vol. 30 No. 2.

1969. *Underdevelopment in Spanish America*. George, Allen & Unwin. London.

Gunder Frank, A., 1963. The varieties of land reform. *Monthly Review* Vol. 15 No. 12.

1964. On the mechanisms of imperialism: the case of Brazil. *Monthly Review* Vol. 16 No. 5.

1967. *Capitalism and underdevelopment in Latin America*. Monthly Review Press. New York.

Halperín, T., 1969. *História contemporanea de América Latina*. Alianza. Madrid.

Hennessey, A., 1978. *The frontier in Latin American history*. Edward Arnold. London.

Hindess, B. & Hirst, P. Q., 1975. *Pre-capitalist modes of production*, pp. 1–20; 221–259. Routledge & Kegan Paul. London.

Hirschman, A., 1963. *Journeys towards progress*. 20th Century Fund. New York.

Hobsbawm, E. J., 1959. *Primitive rebels*. University of Manchester Press.

1969. *Bandits*. Weidenfeld & Nicolson. London.

1969. A case of neo-feudalism: La Convención, Peru. *Journal of Latin American Studies* Vol. 1 No. 1.

1973. Peasants and politics. *Journal of Peasant Studies* Vol. 1 No. 1.

Huizer, G., 1973. *Peasant rebellion in Latin America*. Penguin. Harmondsworth.

Huntington, S., 1968. *Political order in changing societies*. Yale University Press.

Hutchinson, B., 1960. *Mobilidade e trabalho*. Centro Brasileiro de Pesquisas Educacionais – Instituto Nacional de Estudos Pegagógicos. Rio de Janeiro.

Ianni, O., 1961. A contribuição do proletariado agrícola no Brasil. *Revista Brasileira de Estudos Políticos* No. 12.

1964. Political process and economic development. *New Left Review* Nos. 25 & 26.

1968. *O colapso do populismo no Brasil*. Civilizacão Brasileira. Rio de Janeiro.

1970. *Imperialismo y cultura de la violencia en América Latina*. Siglo XXI. Mexico.

1971. *Estado e planejamento no Brasil: 1930–1970*. Civilização Brasileira. Rio de Janeiro.

Incão e Mello, M. C., 1975. *O bóia-fría: acumulação e miséria*. Vozes. Petrópolis.

Jaguaribe, H., 1969. *Desenvolvimento econômico e desenvolvimento político*. Fundo de Cultura. Rio de Janeiro.

Kahil, R., 1973. *Inflation and economic development in Brazil 1946–1963*. Oxford University Press. London.

Katzman, M. T., 1977b. *Cities and frontiers in Brazil: regional dimensions of economic development*. Harvard University Press. Cambridge.

Klein, L., 1976. A nova ordem legal e a redefinição das bases de legitimidade. In eds. L. Klein & M. Figuereido *Legitimidade e coerção no Brasil pos 64*. Forense. Rio de Janeiro.

Kleinpenning, J. M. G., 1975. *The integration and colonization of the Brazilian portion of the Amazon basin*. Institute of Geography and Planning. Nijmegen. Holland.

Kroeber, A. L., 1948. *Anthropology*. Harcourt, Brace. New York.

Laclau, E., 1969. Modos de producción, sistemas economicos y población excedente. *Revista Latinoamericana de Sociologia* No. 2.

1971. Feudalism and capitalism in Latin America. *New Left Review* No. 67.

1975. The specificity of the political. *Economy and Society* Vol. 4 No. 1.

Lafer, C., 1975. *O sistema político brasileiro*. Perspectiva. São Paulo.

Leeds, A., 1964. Brazilian careers and social structure. *American Anthropologist* Vol. 66 No. 6.

Leff, N. H., 1968. *Economic policy-making and development in Brasil 1947–1964*, Wiley. New York.

Lenin, V. I., 1967. *The development of capitalism in Russia*. Progress Publishers. Moscow.

1970. *Imperialism, the highest stage of capitalism*. Progress Publishers. Moscow.

1972. *The state and revolution*. Progress Publishers. Moscow.

Lenski, G., 1966. *Power and privilege: a theory of social stratification*. McGraw Hill. New York.

Linz, J. J., 1964. An authoritarian regime: Spain. In *Cleavages, Ideologies and Party Systems*. E. Allardt & Y. Littumen, eds. Transactions of the Westermarck Society Vol. X. Helsinki.

Luxemburg, R., 1973. *The accumulation of capital*. Monthly Review Press. New York.

Mahar, D. J., 1978. Política do desenvolvimento para a Amazônia: passado e presente. In Baer, Geiger & Haddad eds., *Dimensões do desenvolvimento brasileiro*. Campus. Rio.

1979. *Frontier development policy in Brazil: a study of Amazônia*. Praeger. New York.

Mandel, E., 1968, L'accumulation primitive et l'industrialization du Tiers-Monde. In *En Partant du 'Capital'*. ed. Victor Fay. Anthropos. Paris.

1976. *Late capitalism*. New Left Books. London.

Marcondes, J. V. F., 1963. O estatuto do trabalhador rural e o problema da terra. *Cadernos Brasileiros* 4.

Margolis, M., 1973. *The moving frontier: social and economic change in a southern Brazilian community*. University of Florida Press. Gainesville.

1977. Historical perspectives on frontier agriculture as an adaptive strategy. *American Ethnologist*, Jan.

Marini, R. M., 1969. *Subdesarrollo y revolución*. Siglo XXI. Mexico.

1972. La reforma agraria en América Latina. *Transición al socialismo y experiencia chilena*. CESO: CEREN. Santiago.

Martinez-Alier, J., 1974. Peasants and labourers in southern Spain, Cuba, and highland Peru. *Journal of Peasant Studies* Vol. 1 No. 2.

Martins, A. F., 1962. Alguns aspectos da inquietação trabalhista no campo. *Revista Brasiliense* 40.

Martins, J. de Souza, 1969b. Modernização agrária e industrialização no Brasil. *América Latina* No. 2.
1975. *Capitalismo e tradicionalismo: estudos sobre as contradições da sociedade agrária no Brasil*. Pioneira. São Paulo.
Martins, W., 1955. *Um Brasil diferente*. Anhembi. São Paulo.
Marx, K., 1964. *Pre-capitalist economic formations*. E. Hobsbawm, ed. Lawrence and Wishart. London.
1970. *Capital* Vol. 1. especially book VIII, 'The so-called primitive accumulation'. Lawrence and Wishart. London.
1971. *A contribution to the critique of political economy*. M. Dobb, ed. London.
1972. *18th Brumaire of Louise Bonaparte*. Progress Publishers. Moscow.
1974. *Capital* Vol. III pp. 748–813 & passim. Lawrence and Wishart. London.
1975. *Texts on methods*. T. Carver, ed. Blackwell, Oxford.
Mata, M., Carvalho, E. U. R. & Castro e Silva, M. T., 1973. *Migrações internas no Brasil*. IPES. Rio de Janeiro.
Meireles, J., 1974. Role de l'État dans le dévelopment du capitalisme industriel au Brésil. In *Critique de l'Économie Politique* 16–17. April. Maspero. Paris.
Melhoramentos Norte do Paraná, 1975. *Colonização e desenvolvimento do norte do Paraná*. EDANEE. São Paulo.
Mercadante, P., 1965. *A consciência conservadora no Brasil*. Saga. Rio de Janeiro.
Mesquita, O. V., & Silva, S. T., 1970. Regiões agrícolas do Paraná: uma definição estatística. *Revista Brasileira de Geografia* No. 1.
Messias Junquiera, 1964. *As terras devolutas na reforma agrária*. Empresa gráfica da *Revista dos Tribunais*. São Paulo.
Monbeig, P., 1952. *Pionneurs et planteurs de São Paulo*. Armand Colin. Paris.
1957. Evolução de gêneros de vida rural tradicionais no sudoeste do Brasil. *Novos Estudos de Geografia Humana Brasileira*. DIFEL. São Paulo.
Monteiro, D. T., 1961. Estrutura social e vida econômica em uma área de pequena propriedade e de monocultura. *Revista Brasileira de Estudos Políticos* No. 12. October.
Nelson, M., 1973. *The development of tropical lands: policy issues in Latin America*. Johns Hopkins University Press. Baltimore.
Nicholls, W. H., 1969. The transformation of agriculture in a semi-industrialized country: the case of Brazil. *The Role of Agriculture in Economic Development* (National Bureau for Economic Research), ed. E. Thorbecke. Columbia University Press. New York.
Nicholls, W. H. & Paiva, R. M., 1965a. Structure and productivity of Brazilian agriculture. *Journal of Farm Economics* 47, May.
1965b. Estágio de desenvolvimento técnico da agricultura brasileira. *Revista Brasileira de Economia* No. 3. June–September.
Normano, J. F., 1935. *Brazil: a study of economic types*. University of North Carolina. Chapel Hill.
North, D., 1964. Agriculture in regional economic growth. In *Agriculture in Economic Development*, eds. Eicher & Witt. McGraw Hill. New York.
Nunez Leal, V., 1975. *Coronelismo, enxada e voto*. Alfa-Omega. São Paulo.
Oliveira, F., 1972b. A economia brasileira: crítica a razão dualista. *Estudos CEBRAP* 2. São Paulo.

1976. A produção dos homens: notas sobre a reprodução da população sob o capital. *Estudos* CEBRAP 16. São Paulo.

Paiva, R. M., 1966. Reflexos sobre as tendências da produção, da productividade e dos preços do setor agrícola do Brasil. *Revista Brasileira de Economia* No. 3. June–September.

1968. O mecanismo de autocontrôle no processo de expansão da melhoría técnica da agricultura. *Revista Brasileira de Economia* No. 3. June–September.

1971. Modernização e dualismo tecnológico na agricultura. *Pesquisa e Planejamento* – IPEA Vol. 1 No. 2. December.

Parlmutter, A., 1969. The praetorian state and the praetorian army. *Comparative Politics* Vol. 1. No. 3.

Pastore, J., 1969. *Brasília: a cidade e o homem.* Universidade de São Paulo. São Paulo.

Pereira, L., 1971. *Estudos sobre o Brasil contemporâneo.* Pioneira. São Paulo.

Pereira de Queiroz, M. I., 1967. O sitiante brasileiro e as transformações da sua situação socio-econômica. *Problèmes agraires des Amériques Latines* CNRS. Paris.

1969. *O Mandonismo local na vida política brasileira.* Universidade de São Paulo.

Pinto, L. F., 1976a. É possível controlar a ocupação da Amazônia? *Teoria, debate, informação* No. 1. April.

1976b. *A integração da Amazônia.* Ibid. No. 2. August.

Plaisant, C. A., 1908. *Scenário paranaense – descripção geográfica, política e histórica do estado do Paraná.* Curitiba.

Portela, F., 1979. *Guerra de guerrilhas no Brasil.* Passado y Presente 2. Global. São Paulo.

Porto Tavares, V. & Monteiro, C., 1972. *Colonização dirigida no Brasil.* INPES/IPEA, Rio de Janeiro.

Poulantzas, N., 1968. *Pouvoir politique et classes sociales.* Maspero. Paris.

1973. Marxism and social classes. *New Left Review* No. 78. March–April.

1974a. Internationalization of capitalist relations and the Nation State. *Economy and Society* Vol. 3 No. 2. May.

1974b. *Fascism and dictatorship.* New Left Books. London.

Prado Jr, C., 1962a. *História econômica do Brasil.* Brasiliense. São Paulo.

1962b. Contribuição para a análise da questão agrária no Brasil. *Revista Brasiliense* nos. 28 Mar.–Apr. 1960; 43 June–Sept. 1962.

1966. *A revolução brasileira.* Brasiliense. São Paulo.

Prebayle, M. R., 1967. *Geographie rurale des nouvelles colonies du Haut Uruguay (Rio Grande do Sul).* Bulletin de l'Association de géographes français, publié avec le concours du CNRS nos. 350/351. Paris.

Preobrazensky, E. A., 1965. *The new economics.* Clarendon Press. Oxford.

R.B.G. (*Revista Brasileira de Geografia* 1970 – *Cidade e região no sudoeste paranaense*), 1970. Ano. 32 No. 2. separata.

Redfield, 1960. *The little community.* University of Chicago Press, Chicago.

Reichel-Dolmatoff, G., 1971. *Amazonian cosmos.* University of Chicago Press. Chicago.

Rey, P. P., 1973. *Les alliances des classes.* Maspero. Paris.

Ribeiro, D., 1970. *As Américas e a civilização.* Civilização Brasileira. Rio de Janeiro.

Roche, J., 1959. *La colonisation allemande et le Rio Grande do Sul.* Université de Paris (Institute des Hautes Études). Paris.

1968. *A colonização alemã no Espirito Santo.* DIFEL. São Paulo.

Ronfeldt, D., 1973. *Atencingo: the politics of agrarian struggle in a Mexican ejido.* Stanford University Press. Stanford.

Sá, F., 1970. Transamazônica só, não resolve. *Indústria e desenvolvimento* Vol. III No. 9. São Paulo.

Sachs, I., 1969. *Capitalismo de Estado e subdesenvolvimento.* Vozes. Petrópolis.

Saldanha, N., 1964. A Revolução e seus aspectos político e jurídico. *Revista Brasileira de Estudos Políticos.* July.

Samaniego, C. & Sorj, B., 1974. Articulaciones de modos de producción y campesinados en América Latina. CISEPA (Centro de Investigaciones Sociales, Económicos e Antropológicas) Pontífica Universidad Católica del Perú. Serie: Publicaciones prévias No. 1. Nov.

Santos, R., 1968. O equilíbrio da firma aviadora e a sua significação econômica. *Pará Desenvolvimento* 3:7–30.

Schattan, S., 1961. Estrutura econômica da agricultura paulista. *Revista Brasiliense* 37.

Schmitter, P. C., 1971. *Interest conflict and political change in Brazil.* Stanford University Press. Stanford.

Schuh, E., 1969. Comment on Nicholls' article in ed. E. Thorbecke, *The Role of Agriculture in Economic Development.* Columbia University Press. New York.

Schwartzman, S., 1975. *São Paulo e o Estado Nacional.* DIFEL. São Paulo.

Sen, A. K., 1966. Peasants and dualism with or without surplus labour. *Journal of Political Economy* 74.

Shanin, T., ed., 1971. *Peasants and peasant societies.* Penguin, Harmondsworth.

Sigaud, L., 1973. Trabalho e tempo histórico entre os proletários rurais. *Revista da Administração das Empresas* Vol. 3. Fundação Getúlio Vargas. São Paulo.

Silva, J. G., 1971. *A reforma agrária no Brasil.* Zahar. Rio de Janeiro.

Simonsen, R., 1973. *A evolução industrial do Brasil.* Nacional. São Paulo.

Singer, P., 1963. Agricultura na bacia Paraná-Uruguai. *Revista Brasileira de Ciências Sociais.* July.

1965. Ciclos de conjuntura em economias sub-desenvolvidas. *Revista Civilização Brasileira* No. 2.

1968. *Desenvolvimento econômico sob a prisma da evolução urbana.* Nacional. São Paulo.

1973. Migrações internas. *Economia política de urbanização* CEBRAP. São Paulo.

Skidmore, T., 1967. *Politics in Brazil: an experiment in democracy.* Oxford University Press. London.

Smith, T. L., 1963. *Brazil: people and institutions.* Louisiana State University Press. Baton Rouge.

Soares, R. N. G., 1959. *Interpretação do valor da terra agrícola.* São Paulo.

Stavenhagen, R., 1965. Classes, colonialism and acculturation. *Studies in Comparative International Development.* 1. No. 6.

1968. Seven fallacies about Latin America. In *Latin America: reform or revolution?*, eds. Petras & Zeitlin. Fawcett Premier. Greenwich.

1971. *Las classes sociales en las sociedades agrárias.* Siglo XXI. México.

Stefannini, L. L., 1977. *A propriedade no direito agrário*. Livraria dos Tribunais. São Paulo.

Stepan, A. C., 1971. *The military in politics: changing patterns in Brazil*. Princeton University Press. Princeton.

1973, ed. *Authoritarian Brazil*. Yale University Press. New Haven.

Sunkel, O. & Paz. P., 1970. *El subdesarrollo latinoamericano y la teoria del desarrollo*. ILPES. Mexico.

Taylor, G. R., ed., 1972. *The Turner thesis concerning the role of the frontier in American history*, D. C. Heath. Boston.

Tepicht, J., 1969. Les complexités de l'économie paysanne. *Information sur les sciences sociales*. Conseil International des Sciences Sociales.

Trotsky, L., 1965. *The history of the Russian Revolution*. Gollancz. London.

Tupiassú, A., 1965. Condições socio-estruturais e participação política na Amazônia. Rio de Janeiro.

1968. O processo demográfico da Amazônia. IDESP-SEES. Série Documentos Breves. Aug.

Turner, F. J., 1921. *The frontier in American history*. Henry Holt. New York.

Velho, O. G., 1969. O conceito do camponês e sua aplicação a análise do meio rural brasileiro. *América Latina* No. 1. Jan.–Mar.

1972. *Frentes de expansão e estrutura agrária*. Zahar. Rio de Janeiro.

1976. *Capitalismo autoritário e campesinato*. DIFEL. São Paulo.

Veliz, C., ed., 1965. *Obstacles to change in Latin America*. Oxford University Press. London.

Vianna Moog, 1955. *Bandeirantes e pioneiros*. Globo. Rio de Janeiro.

Vilaça, M. V. & Albuquerque, R. C., 1965. *Coronel, coronéis*. Tempo Brasileiro. Rio de Janeiro.

Vinhas de Queiroz, M., 1966. *Messianismo e conflito social*. Civilização Brasileira. Rio de Janeiro.

Wagley, C., ed., 1974. *Man in the Amazon*. University of Florida Press. Gainesville.

Waibel, L. H., 1955. As zonas pioneiras do Brasil. *Revista Brasileira de Geografia* No. 4. Oct.–Dec.

Weber, M., 1957. *Economia y sociedad*. Fondo de Cultura, Mexico. Vol. II.

1970. *From Max Weber* trans. H. H. Gerth & C. W. Mills. Routledge. London.

Weffort, F., 1965. Raízes sociais do populismo em São Paulo. *Revista Civilização Brasileira* No. 2. May–June.

1966. State and mass in Brazil. *Studies in Comparative International Development* Vol. II.

Westphalen, C. M., 1969. *História do Paraná* Vol. I. Grafipar. Curitiba.

Wheare, K. C., 1953. *Federal government*. Oxford University Press. London.

Wolf, E., 1966. *Peasants*. Prentice Hall. New York.

1969. *Peasant wars of the 20th century*. Harper & Row. New York.

Wolpe, H., 1972. Capitalism and cheap labour-power in South Africa: from separation to apartheid. *Economy and Society* Vol. 1 No. 4.

1975. The theory of internal colonialism – the South African case. *Beyond the Sociology of Development*, ed., I. Oxaal. London.

Zibetti, D. W., 1969. *Legislação agrária brasileira*. Distribuidora de Jornais, Revistas, Livros Ltda. São Paulo.

Interviews

Much of the information contained in the book was gathered through interviews. Listed below are the principal informants, many of whom were interviewed on several separate occasions (and some over a period of some years). They are classified according to geographical location, and, where applicable, the institution where they work (or used to work). Thus, no attempt is made to list them in 'order of importance', nor has it been possible to include everyone who provided insight into the process of frontier expansion in Brazil, and so helped to shape the material of the book. Please note that none of these informants is responsible for the views expressed in this book.

RIO DE JANEIRO

Personnel of the following institutions:
Centro Latino-Americano de Pesquisas em Ciencias Sociais.
Fundação Getúlio Vargas (esp. Aspásia Alcantara Camargo).
IBGE (esp. Roberto Lobato Correa, and M. Magdalena Vieira Pinto in the Cartographic section).
IICA (Instituto Interamericano de Ciencias Agrícolas, esp. Juan Diaz Bordenave, then Director).
INCRA (esp. Osmar Fávaro, Angela Moraes Neves, Dryden Castro de Arezzo, Oscar Teixeira Machado, and Julio Ramires).
IPEA (Instituto de Pesquisa em Economia Aplicada, esp. Aníbal Villela, then director, Vania Porto Tavares, and Claudio M. Castro).
IUPERJ (esp. Machado).
Ministério da Fazenda
Museu Nacional (esp. Octavio Guilherme Velho, director of the post-graduate programme in social anthropology, and Ligia Sigaud).
and also Eurico de Lima Figueiredo, Orde Mordon, Luis Jorge Wernek Vianna, Lucia Klein and Paulo Simões Correa.

SÃO PAULO

Personnel of the following institutions:
BADESP (Banco de Desenvolvimento de São Paulo, esp. Roberto Cano de Arruda).
CEBRAP (Centro Brasileiro de Análise e Planejamento, esp. Octavio Ianni, Francisco de Oliveira, Paulo Singer, Pedro Calil Padis, Frederico Mazzuchelli, Regis de Castro Andrade, Felipe Reichstul, and Fernando Henrique Cardoso, the director of the Centre).
CESP (Companhia Eléctrica de São Paulo, esp. Guilherme Junquiera).
Comissão Interestadual da Bacia Paraná-Uruguai, esp. Raul Czarny.
Faculdade de Ciências Sociais, Universidade Federal de São Paulo, esp. José de Souza Martins and Maria Isaura Pereira de Queiroz.
Faculdade de Ciências Sociais, Universidade de Campinas, esp. Manoel Berlinck, the director.

Departamento de Imigração e Colonização, esp. Antônio Jordão Neto.
IBGE, esp. Dona Maria Luisa.
Instituto de Economia Rural, esp. Dr Thomazini.
IPE (Instituto de Pesquisa em Economia, of USP, esp. Douglas Graham, Egas Nunez, and José Pastore).
Secretaria de Economia e Planejamento, esp. Roberto Amaral, then Secretary.
and also Sebastião Advincula da Cunha (formerly of ASPLAN) and Dr Amorin (representative of the government of Mato Grosso).
In Piricicaba (São Paulo)
Personnel of the following institution:
Escola Superior da Agricultura, Luis de Queiroz, esp. José Molina & Maria Ignes Guerra Molina of the Escola de Agronomia, and Armando Barros de Castro of the Departamento de Economia.

PARANÁ

In Curitiba (the capital) personnel of the following institutions:
BADEP (Banco de Desenvolvimento do Paraná, esp. Marcus Pinheiro Machado, Ruy Neves Ribas, and Michael Wildberg).
Banco de Brasil (esp. Dr Ratto).
DATM (Departamento de Assistência Técnica aos Municípios, esp. João Augusto Fleury Rocha).
DEE (Departamento de Estatística Estadual, esp. Constantino Comninos).
DGTC (esp. Coronel Clovis da Cunha Viana, then director – formerly of GETSOP; Drs Levi Lustoza and Muniz; Dr Raul; Heloise Barthelmes of the Divisão de Geografia).
FPCI (Fundação Paranaense de Imigração e Colonização, esp. Dr Almir Miro Carneiro, Cid Loyolla, Fonseca).
GETSOP (esp. Coronel José Arnaldo Teixera Bollina).
IBDF (Instituto Brasileiro de Desenvolvimento Florestal, esp. Dr Musmauer).
IBGE
INDA (Instituto Nacional de Desenvolvimento Agrário, before substitution by INCRA, esp. José Orontes Pires).
INCRA (esp. Paulo Somers and Maria Angela Somers, João Oracy Marx and Germano Foster, Clovis, and Aroldo Moletta, Coordenador do Setor do Cadastramento).
Patrimonio da União
Secretaria da Agricultura (esp. Djalma Burico Faraco, then Secretary for the Interior).
Universidade Federal do Paraná (esp. Cecilia Maria Westphalen of the Departamento da História; Brasil Pinheiro Machado, director of the Faculdade de Filosofia; and General Luis Carlos Tourinho, director of the Faculdade de Engenharia).
Interviews were also conducted with:
Bento Munhoz da Rocha (state governor 1951–56).
Moisés Lupión (state governor 1946–51; 1956–61).
Ney Braga (state governor 1961–66).
Ivo Thomasoni (state deputy)
José Burigo (formerly director of the DGTC and of BADEP).

Ciro Maracini (formerly of INDA).

Rubens Amaral; Luis Gastão Alencar; Luis Dalcanale and his son Roger Dalcanale, Carlos Ruaro.

In Francisco Beltrão (in the west of Paraná) interviews were conducted with the following persons:

Antonio de Paiva Cantello (prefect 1965–69); Deny Schwartz (prefect 1968–73); Simão Brugnago Neto (GETSOP representative); Carlos Henrique da Silva Novita (ACARPA representative); Inivaldo Martini (agronomist); Oswaldo Aranha (agronomist and former INDA representative in the west); Olivio Cardoso Poletto (former CANGO employee); Dr Walter Pecoits; the President and officials of the Cooperativa Mista de Francisco Beltrão; peasants from the Linha São Miguel; peasants from the region attending a Curso de Liderança Sindical sponsored by ASSESOAR.

In Pato Branco (in the west of Paraná) interviews were conducted with the following persons:

Alberto Cattani (prefect); Milton Popija (agronomist); peasants of Pato Branco, Vitorino and Mariópolis.

In Pérola do Oeste (in the west of Paraná):

Octavio de Mato, the prefect, and peasants of the area, esp. Antonio Soberai, Tercilio Meotti and Ricardo Bonadimann of the Bela Vista district.

In Cascavél (in the west of Paraná) interviews were conducted with personnel of:

INCRA (esp. Coronel Juarez Costa de Albuquerque, Pacheco, Aristides de Oliveira Coelho, Eugenio Armilfo Ritter, Hiroki Shigueru).

In Foz do Iguaçu (in the west of Paraná) interviews were conducted with the following persons:

Antonio Ayres de Aguirra (of the Land Registry Office); Fernando Salinet (Notary Public); Saulo Ferreira (Public Prosecutor); Antonio Bordin, and Ignacio Rangel Batista (founder of Cascavél in 1931).

Note: In the west further interviews were run in Medianeira, Planalto, Capanema and Chopinzinho, which cannot be recorded here. Frederico de Carli of Chopinzinho made many of these interviews possible.

RIO GRANDE DO SUL

In Porto Alegre (the capital) interviews were conducted with the personnel of the following institutions:

IBGE (esp. João Octavio Felicio, José Alberto Moreno – director of the Unidade de Geografia e Cartografia, and in particular, Gervásio Rodrigo Neves, formerly of IGRA and INCRA).

IEPE (Instituto de Estudos e Pesquisas Econômicos, esp. Mauricio Filcther, then director, Laudelino Medeiros, and Fachel).

IICA (esp. Samuel Miragem, then director).

INCRA (esp. Gilberto Bampi, André Forster, Egon Krakhecke, Milton Bins, and Edgar Henrique Klever).

Secretaria da Agricultura (esp. Dr Israel Farrapo).

SUDESUL (esp. Antonio Drumond and Luis Carlos Zancau).

MATO GROSSO

Interviews were conducted principally in Campo Grande, Dourados, and the small towns of the south with:

Joaquim Alan Kardec Adrien
Dolor Ferreira de Andrade
Waldemiro Muller do Amaral
Carlito Batistoti
Camillo Boni
Bernardes Bais Neto
Coronel Clovis Rodriquez Barbosa
Magno Coelho
Dr Antonio Mendes Canale
Alfonso Simões Correa
Fernando Correa da Costa
José Elias
Sueli de Faro Freire
Candido Fernandes
Lourival Fagundes
Arnaldo Estevão de Figueiredo
Vitor-Diogo Guimarães
Osman Gutierrez
Bernardo Martins Lindoso
Eduardo Metelli
Demostenes Martins
Plinio Martins
Dr Paulo Machado
Rene Muzzi
Humberto Neder
Tom Owens
Manoel Rezende
Alceu Sanchez
Jorge Siufi
Antonio Tonani
Dr Josefino 'Pepito' Ujacow
Dr Pindura Veloso

BELO HORIZONTE

Personnel of the following institutions:

CEDEPLAR (Centro de Desenvolvimento e Planejamento Regional, esp. José Alberto Magno de Carvalho, the director, and Charles Wood and Marianne Schmink).

Universidade Federal de Minas Gerais, esp. Malori José Pompermaier and Bernardo Sorj of the Departamento de Ciência Politica.

BRASILIA

Personnel of the following institutions:

INCRA

Universidade Federal de Brasília, esp. Roberto Cardoso de Oliveira, director of the Faculdade de Ciências Sociais.

PARÁ

In Belém (the capital) personnel of the following institutions:

BASA (Banco da Amazônia, esp. Jaime Bevilacqua, journalist)

IDESP (Instituto de desenvolvimento econômico-social do Pará).

INCRA (esp. Elias Seffer, the former director).

ITERPA (Instituto de Terras do Pará, esp. General Linhares de Paiva, the former director, and Paulo Lamarão).

O Liberal, esp. Lucio Flávio Pinto, journalist specialising in Amazon region.

Museu Goeldi, esp. Roberto Cortés.

NAEA (Núcleo de Altos Estudos Amazônicos of the Universidade Federal do Pará, esp. Amilcar Tupiassú, Stephen Bunker, Roberto Santos, Paulo Cal).
SAGRI
SUDAM, esp. Maria Teresa of the Departamento de Recursos Humanos.
Also interviews with lawyers specialising in land law and land disputes, esp. José Carlos Castro, Rui Barata, Luis de Lima Stefannini, Sirotheau Correa, Delmiro dos Santos and João Nunez.
Interviews with local and national entrepreneurs, esp. Irapuan Sales Filho, Henrique Osaki, Carlos Galera, and Carlos Moreira.
And with Benedito Monteiro and Angenor Penna de Carvalho (former director of the Faculdade de Engenharia).
In Paragominas (south of Belém) personnel of:
INCRA, esp. Eliel Gomes da Silva, director of the Projeto Fundiário
Prefecture, esp. José Alberto and Cristino.
In Conceição do Araguaia (south of Pará) interviews with the following persons:
Giovanni Queiroz (prefect); Sergio Guimarães (land lawyer); Paulo Botelho de Almeida Prado (lawyer for the local peasants' union); Frei Henrique (priest); Maria de Conceição Quinteiro (principal of CEBRAP research team).

Index

CAMBRIDGE LATIN AMERICAN STUDIES

1 Simon Collier. *Ideas and Politics of Chilean Independence, 1808–1833*
2 Michael P. Costeloe. *Church Wealth in Mexico: A study of the Juzgado de Campellanias in the Archbishopric of Mexico, 1800–1856*
3 Peter Calvert. *The Mexican Revolution 1910–1914: The Diplomacy of Anglo-American Conflict*
4 Richard Graham. *Britain and the Onset of Modernization in Brazil, 1850–1914*
5 Herbert S. Klein. *Parties and Political Change in Bolivia, 1880–1952*
6 Leslie Bethell. *The Abolition of the Brazilian Slave Trade: Britain, Brazil and the Slave Trade Question, 1807–1869*
7 David Barkin and Timothy King. *Regional Economic Development: The River Basin Approach in Mexico*
8 Celso Furtado. *Economic Development of Latin America: Historical Background and Contemporary Problems:* (second edition)
9 William Paul McGreevey. *An Economic History of Colombia, 1845–1930*
10 D. A. Brading. *Miners and Merchants in Bourbon Mexico, 1763–1810*
11 Jan Bazant. *Alienation of Church Wealth in Mexico: Social and Economic Aspects of the Liberal Revolution, 1856–1875*
12 Brian R. Hamnett. *Politics and Trade in Southern Mexico, 1750–1821*
13 J. Valerie Fifer. *Bolivia: Land, Location, and Politics since 1825*
14 Peter Gerhard. *A Guide to the Historical Geography of New Spain*
15 P. J. Bakewell. *Silver Mining and Society in Colonial Mexico, Zacatecas 1564–1700*
16 Kenneth R. Maxwell. *Conflicts and Conspiracies: Brazil and Portugal, 1750–1808*
17 Verena Martinez-Alier. *Marriage, Class and Colour in Nineteenth-Century Cuba: A Study of Racial Attitudes and Sexual Values in a Slave Society*
18 Tulio Halperin-Donghi. *Politics, Economics and Society in Argentina in the Revolutionary Period*
19 David Rock. *Politics in Argentina 1890–1930: the Rise and Fall of Radicalism*
20 Mario Gongora. *Studies in the Colonial History of Spanish America*
21 Arnold J. Bauer. *Chilean Rural Society from the Spanish Conquest to 1930*
22 James Lockhart and Enrique Otte. *Letters and People of the Spanish Indies: The Sixteenth Century*
23 Leslie B. Rout, Jr. *The African Experience in Spanish America: 1502 to the Present Day*
24 Jean A. Meyer. *The Cristero Rebellion: The Mexican People between Church and State, 1926–1929*
25 Stefan De Vylder. *Allende's Chile: The Political Economy of the Rise and Fall of the Unidad Popular*

26 Kenneth Duncan and Ian Rutledge, with the collaboration of Colin Harding. *Land and Labour in Latin America: Essays on the Development of Agrarian Capitalism in the Nineteenth and Twentieth Centuries*

27 Guillermo Lora, edited by Laurence Whitehead. *A History of the Bolivian Labour Movement, 1848–1971*

28 Victor Nunes Leal, translated by June Henfrey. *Coronelismo: The Municipality and Representative Government in Brazil*

29 Anthony Hall. *Drought and Irrigation in North-east Brazil*

30 S. M. Socolow. *The Merchants of Buenos Aires 1778–1810: Family and Commerce*

31 Charles F. Nunn. *Foreign Immigrants in Early Bourbon Mexico, 1700–1760*

32 D. A. Brading. *Haciendas and Ranchos in the Mexican Bajio*

33 Billie R. DeWalt. *Modernization in a Mexican Ejido: A Study in Economic Adaptation.*

34 David Nicholls. *From Dessalines to Duvalier: Race, Colour and National Independence in Haiti*

35 Jonathan C. Brown. *A Socioeconomic History of Argentina, 1776–1860*

36 Marco Palacios. *Coffee in Colombia 1850–1970: An Economic, Social and Political History*

37 David Murray. *Odious Commerce: Britain, Spain and the Abolition of the Cuban Slave Trade*

38 D. A. Brading (ed). *Caudillo and Peasant in the Mexican Revolution*